The Elan V<

Life in the Elan Valley was not so very different from today. There were reports of wives leaving to live with another partner. Men beating their wives, children going hungry because the fathers kept their hard earned cash for their own requirements. Stealing fighting and quarrelling etc.

The men worked hard under dangerous and difficult conditions but we must also remember the women and children, they, too, experienced hard and difficult times. There were times though when they could forget their hardships such as on May Fair Day in Rhayader. A time for all the family to enjoy themselves, rich and poor, young and old alike. Those from outlying farms walked or came in on horseback. For some it was the only day in the year when they met other folk to exchange news, etc. There being no telephone, radio or TV. There is no electricity in the valley today except that produced by generators.

The Atkinson family enjoying a day at the May Fair.

To my husband, Keith, for his patience and support during the writing of this book.

Foreword

Until now, the only book published about the Elan Valley Water Scheme was that written by Thomas Barclay, a member of the Water Committee, almost a century ago. Not unnaturally, that concentrated on the search for the site, the political battle to abstract the water and the building of the dams.

Now, Rita Morton – who has already earned our gratitude for rescuing so many of the records of the time from destruction – has produced a totally different, and quite fascinating, account which gives a remarkable insight into the conditions of the men living and working at a building site a century ago. But, hard though those conditions seem to us today, we also get sudden glimpses of compassion, and forethought for the workmen, by the Water Committee. For providing this new view of the Elan Valley Works, in such readable form, Rita Morton has again earned our thanks.

Denis Martineau

Contents

Chapter One
Introduction — 1

Chapter Two
The Elan Village — 9

Chapter Three
The Engineers — 57

Chapter Four
Characters of the Valley — 61

Chapter Five
The Building of the Dams — 70

Chapter Six
Expenses incurred during building — 80

Chapter Seven
The Opening — 82

Chapter Eight
What is Water? — 86

Chapter Nine
The Walk — 90

CHAPTER ONE

Introduction

FRESH running water supplied to our homes today is taken for granted. This was not so almost a hundred years ago.

The supply of water which Birmingham receives is due mainly to a group of men led by that wonderful man Joseph Chamberlain. Credit must also be given to Alderman Avery, the first Chairman of the Water Committee. He was followed by Sir Thomas Martineau who was Chairman of the Committee which advised on the adoption of the Welsh scheme and who bore the brunt of the Parliamentary campaign. This arduous task was largely the cause of a breakdown of his health. He died twelve months after the Act was passed.

Joseph Chamberlain came from London to Birmingham in 1854 to represent his father's interest in his uncle's business. Joseph was a perfectionist and a good businessman. He took great interest in Birmingham's social and cultural life and was elected to the Town Council in 1869, becoming Mayor in 1873. He relinquished his Mayoral position and became a Member of Parliament in 1876.

Joseph could see great opportunities for reform and improvement and took them; one being the great Slum clearance (which is where Corporation Street and New Street are now). It was during his term on the Council that many other great improvements came to fruition; the main one – the building of the Elan Valley Dams, near Rhayader in mid-Wales.

Birmingham purchased 71 square miles of Elan Valley so that its citizens could lay claim to all the rain that fell within that area. An average of 70 inches of rain falls in the valley each year.

The cost of this wonderful feat of engineering, the building of the dams, the aqueduct and Frankley Reservoir was £6,000,000 and took twelve years to complete. It was all built by hand, there being no J.C.Bs. etc. It was men with picks and shovels and sticks of dynamite. The navvy was paid 4 pence an hour and his working week was 60 hours. At that time there was a recession, railways and canals had been completed and men came to the Valley looking for work. In those days if you didn't work you didn't eat.

Mr. Joseph Chamberlain

Mr. A. Lees, Secretary of the Water Committee

TO THE CHAIRMAN AND MEMBERS OF THE WATER (ELAN SUPPLY) COMMITTEE
Council Inspection of Elan Valley Works, June 19th 1900

The following is a timed copy of the proceedings throughout:
    Corporation train at Elan Junction 10.55 a.m.   Arrive Public Hall
    Special Train arrive Rhayader 11.05 a.m.   Commence Luncheon 12.00 noon
    Arrive Caban Coch   Speaking commences 12.40 p.m.
Speeches as follows:
    Chairman – "The Queen"
    Lord Mayor to thank Water Committee
    Chairman (a) to reply   (b) to refer to stone-laying   (c) to propose Mr. Mansergh's health
    Mr. Mansergh to reply

Leave Public Hall 1.05 p.m. Leave Caban 1.15 p.m. Arrive Craig Goch 1.45 p.m. Leave Craig Goch 2.15 p.m. Arrive Pen-y-Garreg. Leave Pen-y-Garreg. Arrive Garreg-Ddu. Stone-laying and photograph Leave Garreg-Ddu sharp 4.00 p.m. Arrive Caban 4.20 p.m. Leave Caban 4.40 p.m. Arrive Public Hall 4.5 p.m. Tea – Chairman to make announcement about photographs. Leave Public Hall 5.25 p.m. Leave Caban 5.35 p.m. Arrive Rhayader 6.00 p.m. Leave rhayader 6.10 p.m. Arrive Birmingham 9.22 p.m.

The total Cost of the Inspection was £256.9.8 made up as:
Luncheons, teas, etc. £97.8.5   Railway tickets £77.11.11   Elan Valley Works Charges (seating for trucks, running the party, etc. £38.15.5   Expenses per the Secretary £14.15.7   Plans for Booklet   £10.8.6
Flags £5.3.0   Clerical Assistance £4.10.4   Printing circulars, etc. £4.16.6   Cost of printing booklet £3.0.0

                        Your obedient servant
                              Secretary
July 19th 1900

Minutes about visit to Elan – June 19th 1900

Sir Thomas Martineau

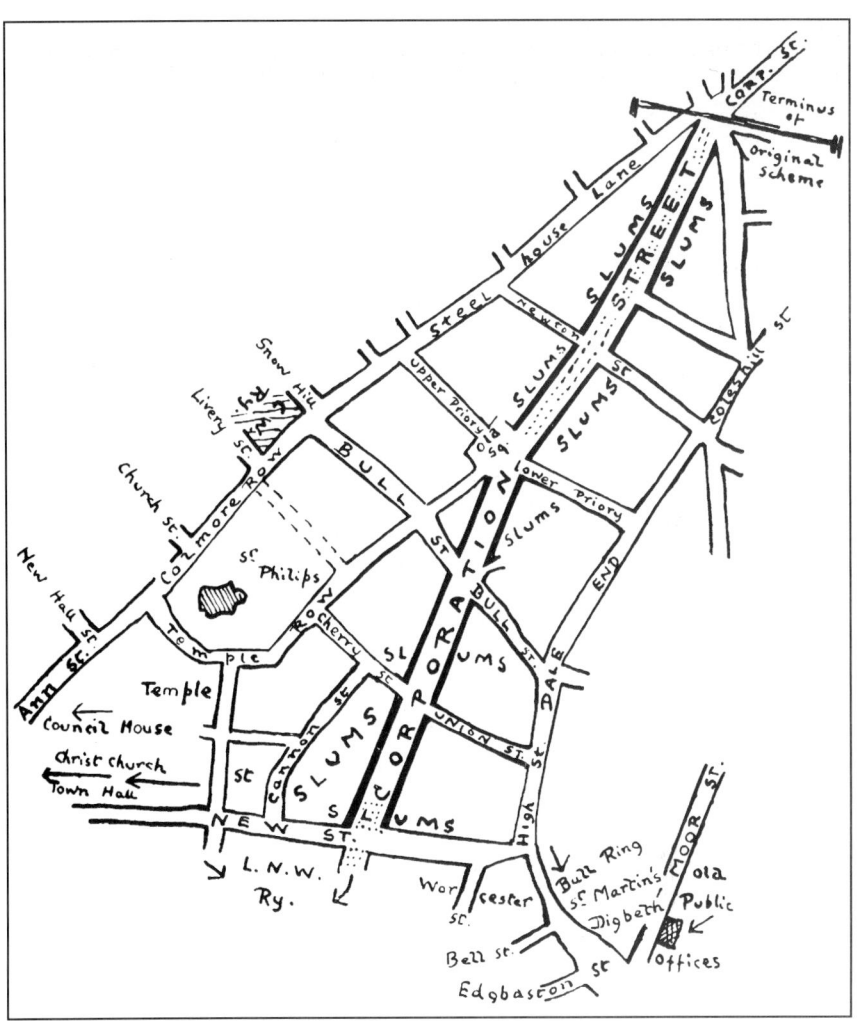

This map clearly shows the Corporation Street Improvement Scheme and all details are omitted. The triangle between Temple Street, Aston Cross, Dale End and Coleshill Street was a densely populated area. The making of Lower Temple Street, the L. & N.W. Station and Stephenson Place had cleared a good many slums. Between Cannon Street and High Street on the north side of New Street was a packed slum area stretching up to Old Square and then along the old Lichfield Street. New Street was pierced at the dotted lines on the map and new thoroughfare driven through to the apex of the triangle of Steelhouse Lane, Stafford Street, Lancaster Street and Aston Street. The original scheme stopped at that point.
The continuation of Corporation Street beyond that point was a later addition to the scheme as planned between 1875 and 1882. Dotted lines in Corporation Street from Old Square northwards indicate the narrow Lichfield Street in the slum area, which the new street displaced.

A view of the slum area of Corporation Street, Birmingham.

The civil engineer in charge of the whole scheme and who planned and designed it was a Mr James Mansergh, who was born in Lancaster. He was a brilliant man, and his work is there for all to see when you visit the Valley and the Dams.

Before any site clearance could begin the area had to be surveyed and James Mansergh sent his two sons and a young man by the name of Eustace Tickell to carry out the surveying. Whilst helping with the surveying Eustace took pen, ink and a sketch pad and sketched the Valley prior to any work commencing on the dams. He was also appointed engineer in charge of the building of the PEN-Y-GARREG Dam (more details of this will be found later in this story). He was an excellent civil engineer and a talented artist.

Mr Yourdi was appointed engineer in charge of the building of the Elan Valley Dams. Yourdi was a strange surname as he came from Cork, his mother being Irish and his Father a Greek who worked for the Greek Government in Cork. Mr Yourdi, a brilliant man, was an expert in concrete and a great disciplinarian. He had to be, with more than 1,500 navvies and engineers under his control. Black was black and white was white as far as he was concerned. Discipline was carried out to the letter without any regard for a person's position. On one occasion one of his junior engineers went to Rhayader and had a drink in one of the public houses there. On his return to the village of Elan he was dismissed by Mr. Yourdi who told him that it was not in keeping with his position to frequent such places. On another occasion the canteen keeper went away for two days at Christmas without permission and on his return he had been replaced.

Mr. Yourdi also complained to the headmaster of the school that children had been found playing see-saws on the trees that had been felled. He wrote "Please make sure that this does not happen again . . ."

In the early 1850s Elan Valley was a small community consisting of cattle and sheep farms with one or two lead mines, with Rhayader being the nearest market town. It was in 1893 that work began on construction of the dams.

A number of buildings had to be demolished including 18 cottages, a church, chapel and school and two residences, "Cwm Elan" and "Nantgwyllt", both of which had connections with the poet Shelley. New buildings were erected in the vicinity to replace those that were destroyed, except Cwm Elan and Nantgwyllt.

It was not an easy task to get this scheme under way. Opposition was very fierce, both in Birmingham and, of course, in Wales.

At the time the dams were being constructed, most members of the Council and its sub-committee were businessmen who received no payment for their services but were loyal and devoted to the people they served, because of their business commitment they usually visited the Elan Valley at weekends staying overnight and returning in time for work on a

Monday morning. They usually paid their own expenses. The train fare being approximately 15s. The journey took about five hours (changing trains at least once and sometimes twice). On arrival at Rhayader they would change from a railway carriage to an open goods wagon. The reason being that a special railway had been laid from Rhayader to the Valley to carry the materials needed in the construction of the dams. Sometimes, depending on the time of their arrival, a pony and trap would be waiting for them.

Council Visit

CHAPTER TWO

The Elan Village

THIS chapter brings to life the village in the 1890s including an insight into the way people worked and lived.

RAILWAY

The Railway was the first phase of work to be done in connecting the mid-Wales section of the Cambrian System with the Elan Valley.

The contractor was Mr. H. Lovatt of Wolverhampton. He received the order on August 16th 1893 and the first section of the railway was completed in July 1894.

Because of problems encountered while building the railway (the importance of safety etc.) Mr. Mansergh advised the Water Committee to build the dams by direct labour and this they did appointing Mr. G. Yourdi engineers-in-charge of the construction.

The Ebbw Vale Steel & Iron Coal Company supplied the rails and fishplates at £5 5s. 0d. per ton for straight rails, £5 8s. 6d. for the curved rails and £6 12s. 6d. per ton for the fishplates. These were required for the railway to be extended to Allt Goch, Garreg-Ddu and the Caban Dam on the Radnorshire side, Crusher Yard, Masons Yard and Cnwch Sidings. There is very little information available with regard to the railway as a great many records have disappeared or been destroyed.

The Oldbury Carriage & Wagon Works supplied 36 end tip wagons at a cost of £361 10s. 0d. and the Ashbury Railway Carriage & Iron Co. 100 side tip wagons £577 10s. 0d.

Six second-hand passenger coaches were purchased from the Great Western Railway at £50 0s. 0d. each. They were supplied without brakes, the railway agreeing to fit a hand brake to one of the coaches at a cost of £3 0s. 0d. bringing the total cost to £303 0s. 0d.

The railway was built specifically for carrying materials for the construction of the dams. Workmen were not allowed to travel on it at all. Some of the men lived in Llanidloes and worked in the Valley from Monday to Saturday, returning home at the weekend. A letter was sent by these men to Mr. Yourdi requesting permission to travel from Rhayader to the valley by train on Mondays, the reason being that the train from Llanidloes arrived too late to allow them time to walk from Rhayader to the valley which caused them to miss the morning shift. Mr. Yourdi considered their request. He checked their statement was in order by timing their departure from Rhayader on Saturday lunchtime and their arrival at Llanidloes. The same check was made in reverse on Monday morning. Permission was not granted.

Locomotives used on the works included: the "Elan", "Claerwen", "Rhiwnant", "Calettwr" and "Coel", all made by Manning, Wardle & Co.; and "Nantgwyllt", "Methan" and "Marchnant" which were made by the Hunslet Engine Co. All were manufactured between 1894 and 1898.

January 20th 1896. It was reported a runaway of a serious character took place on Railway No. 2. There was no loss of life, fortunately, though it resulted in considerable damage to the Corporation's Rolling Stock as well as three trucks belonging to the L. & N.W. Railway Co. One was completely broken up. The ironwork being all that was left of any value.

The Engine "Calettwr" was involved. The estimated damage is put at £120. It is however difficult to say exactly as no bill had been received from the L. & N.W. Railway Co.

Railway between Caban and Pen-y-Garreg.

The Village

The village was built on private ground belonging to Birmingham. The only entrance to the village was by a wire suspension bridge which was built in February 1895 for £19 6s. 7d. There was no public right of way, and only milk was allowed to be delivered or sold on Sundays.

The village had a complete system of sewage, water supply and public lighting. Fire hydrants were fixed onto the water mains, and fire extinguishing appliances were provided at convenient points. In the middle of the village was a small fire station surmounted by a fire bell.

Buildings

All the buildings in the village were constructed of wood. The only brick and stone in the houses being the hearth, the seating for the grates and the chimney flues. Externally, the buildings were weather boarded and internally match-boarded. The space between was lined with coarse felt. The roofs were covered with felt over the boards, and the whole building was then tarred, and in addition the roofs were thoroughly sanded.

Public Buildings

The public buildings comprised a school and mission room, public hall, canteen, hospital, bath house, post office, police station, and adjoining the village, the doss house. There was a hospital with two wards and a hospital for infectious diseases.

The Suspension Bridge and Hospital, Elan Valley

CITY OF BIRMINGHAM WATER DEPARTMENT.

ELAN SUPPLY.

ELAN VALLEY WORKS.

Rules for Bridge-Keeper, Elan Village.

1. To see that no person is allowed to pass the gates who is not a *bona fide* corporation official, hut-keeper, or lodger, or *bona fide* visitor to the huts desirous of seeing friends among the residents, and no tradesman without the necessary pass.

2. To examine each tradesman's van before allowing it to enter the village, with a view to intercept illicit commerce in intoxicants.

3. To open and close the gates when required for the passage of vehicles and foot-passengers.

4. To open the gates for the exit and entrance of the men for the first, second, and fourth quarters.

5. To close the bridge gates every evening at 10 o'clock, excepting Saturday, when the gates will be closed at 11 o'clock.

6. To obey the instructions of the Village Superintendent and of the Resident Engineer.

7. To observe any addition to, or variation of, these rules issued from time to time by the Water Committee.

By order of the Water Committee,

G. N. YOURDI,

Resident Engineer.

A copy of the Rules for the Bridge-Keeper.

Approaching Elan Village from Rhayader Police Station which is the first building on the left.

The Gate-Keeper checking a wagon before entering the village outside the home of the Chitty family.

A gatekeeper was appointed because the Temperance Society and The Band of Hope had complained about the amount of beer that was being drunk and smuggled into the village. A Shebeens (illicit liquor shop) had also been found within the village.

One of the Village streets

A Village street showing the Canteen on the left.

Huts

The huts or houses came in four classes.

1st Class – Ordinary Lodgers

Accommodation at the one end was for the hut keeper, his wife and their family. At the other end there was accommodation for eight lodgers. Midway between the two ends of the hut was the common living room.

2nd Class – Gangers

These were constructed for the overseers and gangers of the workmen and accommodated only one man and his family. Lodgers were permitted in these huts only under exceptional circumstances and by special permit from the resident engineer.

The general custom on public works had been for the ganger to double as hut keeper, meaning the ganger's wife was the landlady of the hut. In many cases she would have been his banker and general provider, and as the ganger himself had power of picking out men to be discharged on account of slackness of work etc. it is easy to see how under such conditions a ganger, seeking his own interests, was able to exercise considerable power over a man in his gang who at the same time was his lodger. In view of this the Water Committee determined at the beginning that no ganger would be allowed to take in lodgers.

3rd Class – for Officials

This accommodation was somewhat more extensive. Most of the houses for officials were erected adjoining the chief's offices and together formed a separate and picturesque group.

4th Class

These comprised of only three rooms, each of which afforded accommodation for married men.

Village Huts

It was usual for the hut-keeper's wife to obtain help with the washing. It was the custom in navvy settlements for women to attach themselves to settlements, generally navvies' widows.

All the huts were to be furnished the same: double beds and bedding for the hut-keeper and his wife; single bed and bedding for the children and lodgers. The bedding comprising of paillasses, flock mattress, pillows, sheets, blankets and quilts.

One of the advantages of the providing of bedding etc. was that it ensured uniformity of comfort. If a particular hut was not patronised the apparent cause was attributed to the hut-keeper and his wife and not to difference in the furnishing this ensured proper control of the hut.

The Village Inspector would visit the hut in order to detect overcrowding, shebeening etc.

Bedsteads for the huts were purchased from Hoyland & Smith Ltd., Soho Works, Heath Street, Springhill, Birmingham.

Beds and paillasses: Mr. G. Lane, Digbeth Bedding Works, Birmingham.

Bed 3'0" x 6'6"   8s. each less 5% discount for cash.

Pair of straw paillasses 6'6" x 3'0"	5s. 6d.
Pair of wool flock mattresses	7s. 0d.
Wool Bolster	1s. 0d.
Pillows	10d.
	14s. 4d.

Pair of straw paillasses 6'6" x 4'6"	7s. 6d.
Wool flock mattress	10s. 0d.
Bolster	1s. 8d.
2 Pillows	1s. 8d.
	£1. 0s. 4d.

Rent for the Huts

Furniture and an allowance of 5cwt. of coal.

Rent 10s. per week.

Rent for each lodger 3s.6d. per week. This provided lodging, tea, potatoes and washing shirt, stockings and trousers.

In May 1894 two tenant farmers who lived in the immediate vicinity of the Caban Coch Works accepted the terms offered by the Water Committee which was to pay 10% interest upon the cost of temporary additions to be made to their houses for workmen.

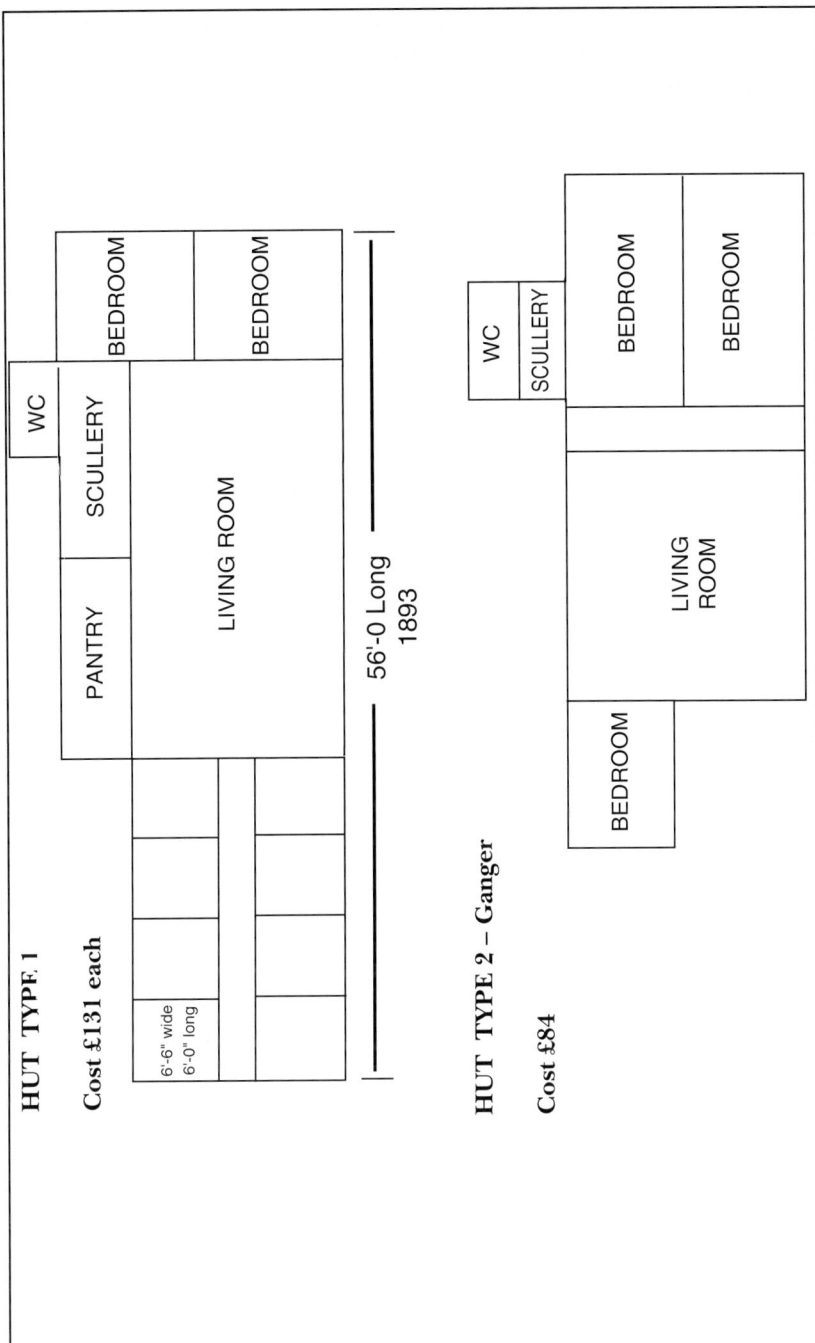

Plans of Huts in the Village

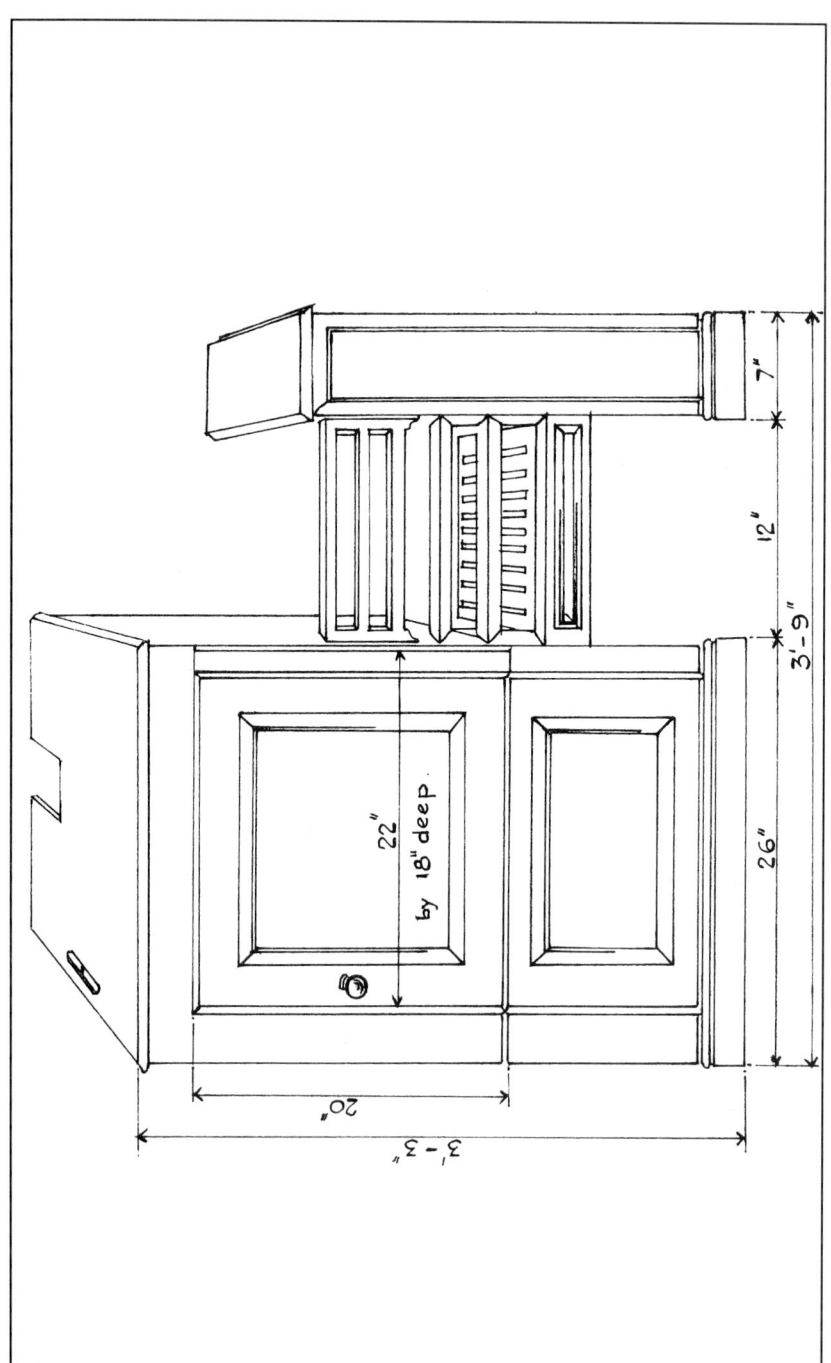

A typical stove in the huts

LANGSTONE WORKS, PENN STREET.
BELMONT ROW.
TELEGRAPHIC ADDRESS:—
RESISTING, BIRMINGHAM.

Birmingham, FEB 17 1894 189

M E A Lees Esq

From Walker & Son.

IRON FOUNDERS &
Fire Proof Safe Manufacturers.

Elan Supply.

Dear Sir,
  We have much pleasure in handing you herewith sketches of Ranges with Ovens reduced to the sizes suggested on the 15th inst:
1 Range 3'·8" with Oven 18½ x 18 x 16" £2·11·6 each free Rhayader
1 Range 3'·4" with Oven 15" x 14" x 14" £2·5·6 each.
These are the same as regards bars and boiler as the 4'·0" one supplied to you on the 29th ult.
  Referring to sample Fenders No 1 and 2 reduced to 9" in height we beg to quote you: No 1 – 5/6 and No 2, 6/6 free Rhayader.
  Trusting these will meet your approbation.
      Faithfully Yours
      Walker & Son

Please note. The price of the 4'·0" range supplied on 29th ult free Rhayader would be £2·16·6.

CITY OF BIRMINGHAM WATER DEPARTMENT
ELAN SUPPLY.

ELAN VALLEY WORKS.

RULES & REGULATIONS
RELATIVE TO

LODGERS
IN
FARM HOUSES AND COTTAGES
In the Occupation of Tenants of the Birmingham Corporation

RULE 1.—The charge to be made by the Tenant for each lodger will be not less than 2/6 per week, or 6d. per night for less than a week. For this, the tenant must undertake to cook whatever may be reasonably required for breakfast, dinner, and supper, and must provide full attendance in all particulars, and, in the case of the weekly lodgers, the washing of the following garments:—
    1 Shirt.   1 pair Socks.   1 pair Stockings.   1 pair Trousers.

RULE 2.—Tenants may supply the lodgers with food by mutual agreement, but lodger are at liberty to provide their own food, which shall be cooked without further charge.

RULE 3.—Each new occupant of a bed shall be provided with clean sheets, pillow and bolster slips, which shall be changed once a fortnight, or oftener if the Village Superintendent (who shall have power to enter at all reasonable hours to inspect the state and condition of the lodgings) so directs.

RULE 4.—No tenant shall take more lodgers than the number notified by the Estate Agent, and each lodger shall be provided with a separate bed.

RULE 5.—The above rules may at any time be added to or varied by the Water Committee. Infringement of any of the above rules, or any other rules hereafter in force, will render the offender liable to have the permission to take lodgers withdrawn.

RULE 6.—A copy of these rules must be posted up in the principal living room of the farm house or cottage.

RULE 7.—The number of lodgers the tenant is allowed to take, in accordance with the above rules and regulations, is

By order of the Water Committee,

STEPHEN W. WILLIAMS
Agent for the Corporation of Birmingham
Elan Estate.

CITY OF BIRMINGHAM WATER DEPARTMENT
ELAN SUPPLY.

ELAN VALLEY WORKS.

RULES & REGULATIONS
RELATIVE TO
WORKMEN'S HUTS

Type 4. MARRIED WORKMEN'S.

RULE 1.—Tenants are not allowed to receive lodgers into their huts, except by special permission from the Resident Engineer, obtained on application through the Village Superintendent.

RULE 2.—The only lamps permitted in the huts are the Defries patent or some other approved safety lamps with metal reservoirs and solid metal or glass bells hung over the funnels. The wooden ceilings over the lamps shall at all times be protected by tiles or sheet iron, with at least 1-inch of air space between the tiles or iron and the ceiling. Under no circumstances shall lamps be cleaned or trimmed by artificial light. These operations shall always be carried out in the daytime, and never within the hut, but always in the separate lamp-room provided behind each hut. No oil whatever shall be stored within the hut, but only in the lamp-room at the back of the hut, and then only in approved, small, properly-closed vessels. Petroleum of the best quality only shall be used.

RULE 3.—The tenant undertakes that the sanitary arrangements connected with the hut shall be kept in a perfect state of cleanliness, and that he will permit no accumulation of dirt, filth, or refuse, about the premises. The hut and its appurtenances shall, at all reasonable times, be open to the inspection of the Village Superintendent, or any other official appointed for the purpose by the Corporation, with a view to see that the foregoing regulations, and any others which from time to time may be in force, are duly observed, and that the property of the Corporation is kept in a proper state of cleanliness and repair.

By order of the Water Committee.

G. N. YOURDI,
RESIDENT ENGINEER.

June 15th, 1894.

GEO. JONES & SON, TOWN HALL PRINTING OFFICES, 57-63, EDMUND STREET, BIRMINGHAM.

The Fire Station

A central fire station was erected in the Elan village at a cost of £52.0s.9d. The fire bell from Mr. Charles Carr of Birmingham was purchased for £7.18s.6d. There were three reports of fires one at the chief's offices, one at Caban Coch and one at the Doss House.

> Birmingham Corporation Waterworks,
> Resident Engineer's Office,
> Elan Valley,
> Rhayader.
> 14th May, 1896.

Dear Sir,

FIRE AT CHIEF OFFICE

A fire occurred about mid-day today at the Chief Office Coal House. It is not exactly known how the outbreak occurred, but it is presumed that someone, possibly a navvy, threw a lighted match outside. A good many men have been paid off today, and I have no doubt they were loitering about waiting for the timekeepers, and in all probability, it was one of these men who inadvertently set fire to the building. The Chief Office Fire Brigade turned out, and extinguished the fire in about five or six minutes. The estimated damage is £4. None of the boards were burnt off, but the greater portion of the interior is very much charred and the building will have to be reconstructed.

Yours very truly,

(signed) G. N. YOURDI.

E. A. Lees, Esq., Birmingham.

The Police Station

The police station was situated on the Radnorshire side of the village. Policing of the village was carried out by the County Constabulary of Brecon, and the works by Radnorshire police. A letter from Mr Edmund Gwynne, Chief Constable of Breconshire, written on 5th October 1896, praised the constable.

The Police Station had two lock-ups plus stables.

A letter dated April 1895 stated that the Water Committee agreed to pay £25 per year to the Chief Constable of Radnorshire to cover his additional travelling and out-of-pocket expenses.

The School

The school which was opened in 1895, also doubled as a mission room being a school on weekdays and a school and service room on a Sunday. The school accommodated for 168 children. At first there was difficulty in getting the children of navvies to attend but when they did, the school received a Government grant. Up until 1903 there were no records of the headmaster, or statistics on numbers, etc. Personnel records show that Mr Upstone was the headmaster in 1903 and, along with his wife, who was a trained assistant, they received a joint salary of £175 per year. There were two assistants, Elizabeth Hill and Jane Evans. Elizabeth, aged 32 years, received £90 per year and Jane, aged 22, received £65 per year. There were three monitors: Grace Jones, Martha Senior and Sarah Jones. Grace received £13 per year, and Martha and Sarah £10 8s. 0d. each. The following shows the cost of furniture and fittings etc.

To the WATER (ELAN SUPPLY) COMMITTEE
Minute 769
(in Orders & Authorities Book),
Furniture and Appliances for School-room

I beg to report that, under the authority to the Chairman to purchase the necessary furniture and appliances for the School, the following purchases have been made:

Midland Educational Co. – School furniture, appliances and books (including bibles and hymn books for Sunday Services)	124 . 12 . 11
Stockley & Sabin – American Organ	15 . 0 . 0
Chamberlain, King & Jones – Mats, etc.	10 . 8 . 6
Fittings made at the works	4 . 0 . 0
	154 . 1 . 5

At the end of each term Mr Upstone would send his report to the Birmingham Education Department.

Bad attendance this term owing to lambing, hay making, measles, influenza and whooping cough, etc.

When attendance was good he would request a day's holiday at the end of term for his pupils and this would be granted.

The examination papers show that school was not so enjoyable as it is today but children would leave the school with the ability to read, write and do simple arithmetic.

A Mr Albert Chitty, who now lives in Birmingham, can remember attending this school and Mr Upstone taking some of the older boys for mental arithmetic. They had to stand in a straight line and when they answered the question correctly they moved to the head of the line.

Mr. Upstone and members of his staff. Mrs. Upstone is standing on her husband's left.

Inside Elan Village School.

NOTICE.

Bathers should not take a Hot Bath above 100 degrees Fahrenheit without Medical advice, and, in such cases, they are requested to see that the Bath Attendant regulates the temperature of the water by the Thermometer.

Bathers must not Trample upon or otherwise Damage, and unnecessarily Dirty the Towels, but, when they have finished Bathing, leave the Door open and deposit the Towels in the Basket placed in the Corridor adjoining.

Any person occupying a Bath Room for a longer period than 30 minutes, is liable to a second charge for admission.

Any person defacing the Mirrors, or otherwise wilfully damaging the Fittings and other articles provided for the public use will be Prosecuted.

A Ticket must be received in exchange for all money paid, which Ticket must be handed to the Attendant, in exchange for the soap and towel.

Purchasers of 3d. Tickets are entitled to two towels.

BY ORDER.

Bath House

For these men to have a hot bath must have been a wonderful experience for them when you consider that they were building dams for the people of Birmingham to receive cold water in their homes.

The cost of building the Bath and Wash House amounted to £275.18s.3d.

The heating apparatus for the baths was supplied by Messrs. Walter Shaw & Co. at a cost of £131.8s.6d. which included Barbers Rooms. Joseph Holding, of Rhayader, was appointed hairdresser in September 1895 and was charged 12s.6d. per week for rent.

CITY OF BIRMINGHAM WATER DEPARTMENT.

ELAN VILLAGE.

BATH & WASH-HOUSE

THE BATH AND WASH-HOUSE
WILL BE
OPENED ON MONDAY, AUGUST 12, 1895.

THE HOURS WILL BE—

FOR MEN:—   TUESDAYS   6 p.m. to 9 p.m.
                  FRIDAYS   6 p.m. to 9 p.m.
                  SATURDAYS 1 p.m. to 9 p.m.
                  SUNDAYS

FOR WOMEN:— WEDNESDAYS 2 p.m. to 5 p.m.

THE CHARGES WILL BE—

FOR A BATH—1st CLASS   - -   **3d.**
Including a Cake of Soap and the use of Two Towels.

FOR A BATH—2nd CLASS   - -   **2d.**
Including a Cake of Soap and the use of One Towel.

FOR LAVATORY -   - - -   **1d.**
Including use of Soap and Towel.

By order of the Water Committee.

G. N. YOURDI,
RESIDENT ENGINEER.

31st JULY, 1895.

BIRMINGHAM CORPORATION WATERWORKS.
ELAN SUPPLY.

RULES
OF THE
ELAN VILLAGE WORKMEN'S CLUB.

1.—All Employees of the Corporation are eligible for membership, and may become members, on application to the President or Secretary, and on agreeing to comply with the Rules.

2.—Each member is at liberty to introduce a friend, who for one week may enjoy the privileges of the Club. An extension of this time may be granted by the Committee.

3.—The officers of the Club are a President (who shall be the Corporation Missioner, ex-officio), and a Secretary, who shall be appointed by the Water Committee.

4.—A Committee of eight shall be chosen annually by the members from their number, to act with the President and Secretary in the management of the Club, particularly to assist in maintaining order, and in promoting amusements and recreation.

5.—Games, Newspapers, &c., and material for letter-writing will be provided. The Games will include Draughts, Halma, Dominoes, and Beanbags.

6.—A Bagatelle Table will be provided for the use of the members, at a charge of 1d. per game, or such other sum as the Club Committee may decide.

Approved by the Water Committee.

E. ANTONY LEES,
Secretary.

Council House, Birmingham,
31st December, 1904.

Public Hall and Recreation Hall

The recreation room was open every night and also during the daytime when the weather was bad. There was a library stocked with books supplied by Birmingham Library; on Sunday afternoons a writing class took place.

On special occasions a lantern show was held for the women and children. The cost of oil for the lamps and tea and buns was 7s. 6d.

Within the complex was a gymnasium where the missioner acted as an instructor to squads of nippers and young men.

Social Council

There were four members from the village with nine standing for election. Those elected were:

William Jones	Foreman Carpenter
James McIsack	Chief Cashier
Drusilla Parks	Nurse
John Pickering	Clerk

When the Recreation Hall was opened there was only one bagatelle table so an application was made for a second table. The reason given was that most of the men did not play very well and took a long time to complete their game. Those who had experience and played well did not always get a game. The second table was purchased for those who were experienced players.

Recreation Room. Note the two bagatelle tables

The Doss House

The Doss House was situated on the Radnorshire side of the river, not in the village itself. It doubled as a working men's hotel or model lodging house and also as a quarantine for the 36 men in occupation.

Accommodation was of a very high standard in comparison with that which the men had during the building of the canals and railways. General Booth, of the Salvation Army, offered to sell to the Water Committee mattresses that had been used in settlements in London but in a letter received from General Booth these were of poor quality (an extract of this letter is shown below).

The Mattress is stuffed with Seaweed, in which animal life cannot exist, while its covering of American Cloth makes it easy to sweep with a brush which is a matter of daily custom, as soon as the men retire. The American Cloth permits the application of Jeye's Purifier, which immediately dries with ventilation.

The accommodation offered in the Salvation Army Shelter was poor in comparison with the Doss House in the Elan Valley. The men slept on bunk beds. Smoking was discouraged though not strictly prohibited. There was no danger of fire as there were no bed clothes. Smoking was allowed in the kitchen. An offer was made to supply a man to take charge of the Doss House and a charge of 2d. per night was suggested. The Water Committee declined the offer and made their own arrangement which were far superior.

The Committee eventually purchased bedsteads from Hoskins of Birmingham priced at 10s. 6d. each and pillows at 10d. each. Blankets were purchased from Witney Blanket Co. at a price of 3s. 2d. each.

Overleaf is a list of furnishing issued to the Doss House and charged through the Doss House Construction Account.

1894		
June 6	Bedsteads	25 . 13 . 0
19	Bedding	67 . 17 . 8
	Lamp	9 . 6
	Oil can	1 . 8
25	Fenders	1 . 3 . 3
July 12	Lamp	9 . 6
14	Knives, etc.	30 . 1 . 10

1894		
July 16	Bolster Slips	1 . 8
Sep. 26	Bolsters	1 . 16 . 0
	Pillows	1 . 8 . 6
	Pillow Slips	11 . 3
Dec. 6	Mangle and Dolly Tub	3 . 0 . 6
Feb. 19	Blankets and Quilts	6 . 6 . 4
March	Oval numbers for beds	4 . 5

LONDON:
36, King Street,
Cheapside.

REGISTERED TELEGRAPHIC ADDRESS:
"BLANKETS, WITNEY."

CHARLES EARLY & CO.,

WITNEY MILLS, OXFORDSHIRE.

Carriage paid on all Parcels of 1 cwt. and upwards. Smaller Parcels paid to London.

TERMS:
2½ % for Cash, or
4 Months' Bill.

QUALITY.	No. 01.		No. 1.		No. 2.		No. 3.		No. 5.		No. 6.		Superfine.		Super.		Witney Twill.		Witney Bath.		Extra Super.	
Price per lb.	12		13¾		15		16½		18¾		20½		21¾		23¾		2/¾		2/10		2/6	
SIZE NUMBER.	Wght.	Price.	Wght.	Price.	Wght.	Price.	Wght.	Price.	Wght.	Price.	Wght.	Price.	Wght.	Price.	Wght.	Price.	Wght.	Price.	Wght.	Price.	Wght.	Price.
8	4.0	4/-	4.0	4/7	4.0	5/-	4.0	5/6	4.0	6/1												
9	5.4	5/3	5.4	6/-	5.8	6/10	5.12	7/11	5.12	8/9	5.12	9/8	6.0	10/10	6.0	11/10			5.0	14/2		
10	7.0	7/-	7.0	8/-	7.0	8/9	7.0	9/7	7.0	10/8	7.0	11/10										
10 Heavy			8.0	9/2	8.0	10/-	8.0	11/-	8.0	12/2			8.0	14/6	8.0	15/10	9.0	18/7	6.12	19/2	9.0	22/6
11	9.0	9/-	9.0	10/4	9.0	11/3	9.0	12/5	9.0	13/8	9.0	15/2										
11 Heavy			10.0	11/6	10.0	12/6	10.0	13/9	10.0	15/2			10.0	18/2	10.0	19/10	11.4	23/2	8.8	24/1	11.4	28/2
12			11.8	13/2	11.8	14/5	11.8	15/10	11.8	17/6	11.8	19/5	12.0	21/9	12.0	23/9	13.4	27/4	10.8	29/9	13.4	33/2
13													14.0	25/4	14.0	27/9	16.4	33/6	12.8	35/5	16.4	40/8
12×13															15.8	30/8			14.0	39/8	18.0	45/-
14															17.0	33/8	19.0	39/2	15.0	42/6	19.0	47/6

PURE MERINO.

	Plain		Twill	
SIZE.	Wght.	Price.	Wght.	Price.
10	6.12	30/8	8.0	36/4
11	8.8	38/7	10.0	45/5
12	10.8	47/8	12.0	54/6
13	12.8	56/9	15.0	68/2
14	15.0	68/2	17.8	79/6

UNBLEACHED.

SIZE.	Super	Extra Super.
10	18/8	22/-
11	23/4	27/6
12	28/-	33/-
13	35/-	41/3

CRIB.

SIZE.	Super.	Witney	Merino
4	2/3	2/9	4/7
5	3/9	4/6	6/8
6	5/4	6/4	9/4
7	7/5	9/1	13/-
8	10/1	12/1	18/1
9			22/9

SCARLET.

SIZE.	No. 0.	No. 2.
10	11/5	18/5
11	14/6	22/9
12	18/3	27/7

FANCY STRIPES.

SIZE.	Windrush.	Super.	Merino.
8	3/9	5/4	8/-
9	5/3	7/5	11/2
10	6/9	9/7	13/10
11		12/9	20/2
12		16/-	

HOUSE FLANNELS.

	Width	Price.
Plain Grey ...	20in.	4½d.
Twill Grey ...	20in.	5½d.
Plain White...	22in.	5½d.
Twill White...	22in.	6½d.
Milled White	24in.	8½d.

GREY.

SIZE......	9	10
Weight...	6.8	8.8
Price......	6/9	8/10

BATH MATS. Various Patterns.

SIZE	5×6	5×7	6×6	6×7
Price	5/2	5/9	5/9	6/9

IRONING BAIZE.

Width	4/4	5/4	6/4	8/4
Price	2/1	2/7	3/-	4/-

April, 1894. All Goods from No. 9 Size and upwards Whipped in Singles.

DOSS HOUSE CARETAKER'S RULES

I. He must see that copies of the "Rules for Lodgers" are hung up in conspicuous places in the Doss House.

II. He must collect from each lodger, each night, the sum of 3d., for which he must issue one of the numbered tickets, provided for the purpose. He must take care that these tickets are issued in strict numerical order.

III. He must not, without special instructions from the Resident Engineer, permit any lodger to remain in the House a second night, unless the lodger produces a "Workman's Doss House Ticket", in which case he will permit the lodger to remain in the House for one week in all.

IV. No lodger employed on the Works must be permitted to remain in the House more than a week, without special instructions from the Resident Engineer.

V. He must pay particular attention to the cleanliness of the House, and of all appertaining to it. To this end, he must see that the bedding of all occupied beds is changed at least once a week, and oftener if necessary.

VI. He must see that all mattresses and pillows are sponged with the disinfectant provided, every morning, in the case of "one night" lodgers, and at the end of the week, or oftener, in the case of lodgers employed on the Works.

VII. He must see that the entire House is thoroughly washed and disinfected at least once a week.

VIII. He must see that all beds and bedding are properly turned down and aired, and the windows in the wards fully opened every morning.

IX. He must see that the chamber utensils are emptied and cleansed immediately after the men leave the wards.

X. He must see that the water closets are kept in a thoroughly clean and sanitary condition.

XI. He must at once report to the Resident Medical Officer in the Elan Village, any case of illness among the lodgers.

XII. He must, each morning, make up his return, in the book provided for the purpose, of Doss House tickets sold the previous evening, and hand it, with the money, to the Village Superintendent, who will sign for it.

XIII. The Doss House Caretaker will, on his carrying out the above rules and generally attending to the good order and administration of the House, have returned to him 1d. for each ticket sold, and will be permitted to live in the House, with his family, rent free, and will be provided with 8 cwt. of coal, 4 lbs. of soap, the necessary disinfectants, and 2 gallons of oil for use in the lodgers' rooms, per week.

By order of the Water Committee,

G. N. YOURDI,

RESIDENT ENGINEER

February, 1895

Inside the Doss House

BIRMINGHAM CORPORATION WATERWORKS.

ELAN VALLEY WORKS.

DOSS HOUSE.
RULES FOR LODGERS.

(1.) The House is intended for the temporary accommodation of persons seeking employment on the Works of the Corporation.

(2.) No Applicant will be admitted unless he consents to have his clothes disinfected and to take a bath.

(3.) The price charged is 3d. each per night, which must be paid on entering. In exchange the Applicant will receive a numbered ticket, entitling him to a night's lodging, with use of clean night-shirt, bed, bed-clothes, and use of the common fire.

(4.) Lodgers are permitted to remain in the House one night only, unless, on the following day, they are successful in obtaining employment on the Works.

(5.) Men who obtain employment are required to remain in the Doss House a week before they will be permitted to take lodgings in the Elan Village, and must each obtain a "Workman's Doss House Ticket" from the time-keeper, and must show it to the care-taker at the Doss House, when applying for lodgings on the second and following nights.

(6.) Provisions can be obtained from the care-taker at the following prices:—

One pint of tea, with milk and sugar	. . .	.	1d.
Potatoes (per meal) - 1d.	Two Red Herrings	.	3½d.
Bacon (per lb.) - 8d.	Bloater (each)	.	1d.
Bread (per lb.) - 1d.	Soup - . .	.	1d.

(7.) Lodgers are permitted to provide their own food and to use the common fire to cook it.

(8.) No one is permitted to bring intoxicating drink of any kind whatever into the House, and anyone infringing this rule will be at once turned out, and, if in employment on the Works, will be paid off, and will have no further chance of being engaged.

(9.) Lodgers must strictly observe the Rules, and must be quiet and orderly in their conduct. The care-taker is authorised to turn out anyone transgressing the Rules.

(10.) Lodgers are requested to help in the maintenance of order, and in the observance of the Rules.

By order of the Water Committee,

G. N. YOURDI,
RESIDENT ENGINEER.

February, 1895.

No. **A 1828**

B. C. W. W.

WORKMAN'S DOSS HOUSE TICKET.

RULE 5—"Men who obtain employment are required to remain in the Doss House a week before they will be permitted to take lodgings in the Elan Village, and must each obtain a 'Workman's Doss House Ticket' from the Timekeeper, and must show it to the Caretaker at the Doss House when applying for lodgings on the second and following nights."

Ticket issued to .... No ....

Gang .... Date ....

(68803) .... Timekeeper.

New Business.

City of Birmingham.
Town Clerk's Office.
September 22 1902.

Water Committee
Elan Supply.

} Report with reference to Minute No

Arson at Elan Valley

I beg to report that I was consulted by your Secretary with respect to the setting on fire of the Corporation Doss. House at Elan Valley on the morning of the 14th instant. The Police had arrested a man named John Williams alias Sullivan who had been employed on the works.

As legal assistance was requested I instructed Mr Reay Nadin, of my office, to attend at Elan Valley and investigate the matter. He found that there was a strong primâ facie case against the prisoner of deliberate arson endangering the lives of some fifty six persons.

The prisoner was brought before the County magistrates at Rhayader on the 17th instant. He strongly denied the allegations against him but after a lengthy hearing was committed to the Brecon Assizes which will be held in November next.

E. O. Smith
Town Clerk

Elan Valley Sick Club

One of the first National Health Schemes was brought into the valley by the Birmingham Water Committee. The date of establishment was August 31st 1894.

Every man had to agree to have sixpence deducted each week from his earnings. This covered himself and his dependants should medical care be required. Boys had to agree to have three pence per week deducted from their earnings.

Workmen earning four pence and upwards per hour were classed as men and those earning less were classed as boys.

Sick pay was 12 shillings per week for 13 weeks and after that six shillings for a further 13 weeks. Appointments had to be made to see the doctor who called once a week.

If the doctor was called out at any other time it was classed as private and a charge was made. A charge was also made for a confinement.

Those joining or leaving during a week pay one pence per day and half pence for boys.

The money was placed in the North & South Wales Bank, Rhayader.

In 1906 6,000 persons had been treated in the outpatients' department from when the work commenced.

Receipts for first quarter	47 . 19 . 0	
Receipts for second quarter	47 . 13 . 1	
	95 . 12 . 1	
Expenditure for first quarter	2 . 17 . 6	Stationery
	15 . 2 . 0	Sick Pay
	14 . 6 . 2	Doctor
second quarter	46 . 10 . 0	Sick Pay
	14 . 19 . 6	Doctor
Cheque book	2 . 1	
Balance in hand	1 . 14 . 10	
	95 . 12 . 1	

The Hospital

The hospital opened in October 1894 with 203 inpatients and 1,320 outpatients being treated from its opening until November 1897. It was situated three miles from Rhayader and seven miles from the works necessitating the siting of first aid posts around the works.

There was accommodation for 18 patients in two wards: a general ward, and one for accidents. Most of the time only one ward was in use with the other two being used occasionally.

It was recommended that for each patient in a medical ward the floor space should not be less than 100 sq. ft. and the cubic space 1,000 cubic ft. In surgical wards and Infectious Hospital the minimum floor space should be 140 sq. ft. with a minimum cubic space 150 cubic feet per head.

A request was also made in March 1895 for a special bedroom for the night nurse as it was impossible for her to sleep during the day. It was built onto the female ward at a cost of £30.00.

In November 1895 a request was made by Nurse Parks for a building to be erected for use as a laundry and drying room. The estimated cost was between £70 and £80. This was agreed upon by the Water Committee. In January 1896 all this changed. Since the proposal to erect the laundry had become known in the village some of the women in the village expressed their willingness to do the washing on reasonable terms. This proposal was accepted.

The Isolation Ward

The head of this ward was Dr. Wilkes his salary being £250 per annum. The ward was situated between the Caban Dam office and Coed-Troed-y-Rhiw-Fach on high ground, and extended in December 1902 to hold 70 beds at a cost of £300.00.

Smallpox

Smallpox was recorded in England and South Wales in 1896, but no outbreak occurred in the village. The whole village with the exception of three were vaccinated and kept in isolation for one week.

There was a smallpox outbreak recorded in May 1903 and many men left the village. The doss house was closed and an isolation period of 14 days was enforced. Five cases were reported one being fatal. The isolation period was changed as can be seen from the letter on page 40.

Typhoid

A report written in October 1896 about the outbreak of Typhoid Fever stated: Up to the present time there have been ten cases, two of whom have died. One letter sent to Birmingham stated that a man had died and his mattress had been destroyed along with his clothes (rags) the question was asked "May his widow have a new mattress?" the reply was "Yes, be sure to get the cheapest quality!".

Resident Engineer's Office
Elan Valley
Rhayader
21st April, 1896

Dear Sir

SMALLPOX EPIDEMIC

After consultation with Dr. Clarke I have reduced the quarantine at the Doss House to one week.

I have also written to Superintendent Palmer to say that men who have a Medical Certificate from either Dr. Clarke or Dr. Richardson are to be admitted straight to the Village. This, of course, does not refer to tramps, but to men of satisfactory character and good personal appearance.

Yours very truly,
(signed) G. N. Yourdi.
p. W.H.P.

E. A. Lees, Esq.,
Birmingham

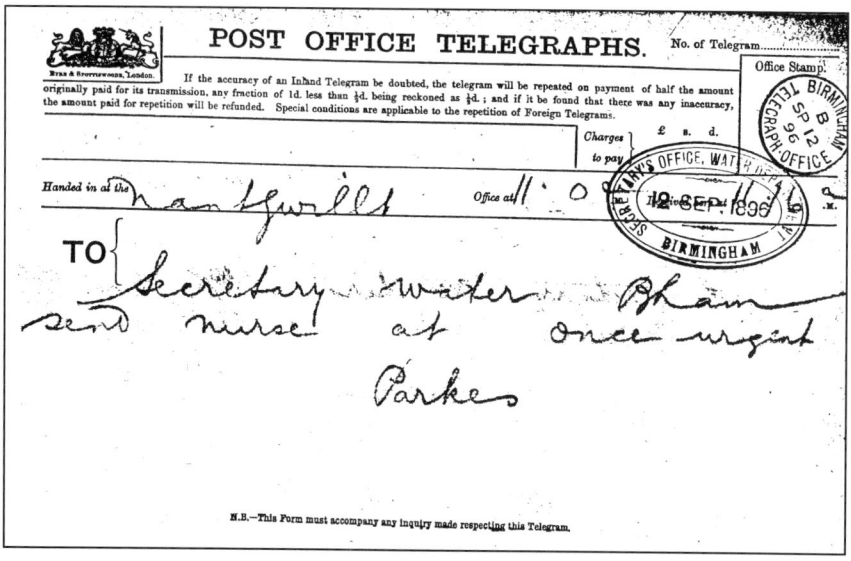

No. 25884.
Birmingham Corporation Water Works.
Resident Engineer's Office
Elan Valley

Epidemic. Rhayader 23rd April 1896.

E. A. Lees, Esq.,
Birmingham

Dear Sir,

After consultation with Dr Clarke I have reduced the quarantine at the Doss House to one week. I have also written to Supt Palmer to say that men who have a medical certificate from either Dr Clarke or Dr Richardson are to be admitted straight to the village. This form,

does not refer to tramps, but to men of satisfactory character and good personal appearance.

Yours very truly,
G. N. Yourdi
p. RAS

BIRMINGHAM CORPORATION WATERWORKS.

ELAN SUPPLY.

ELAN VALLEY WORKS.

ACCIDENTS HOSPITAL.

RULES.

1. The Hospital is intended only for the treatment of Accidents or Surgical Cases. The Employees of the Corporation and their families are alone entitled to be treated. Cases will be admitted to the Hospital, or treated as Out-patients, in the discretion of the Resident Medical Officer.

2. No Medical Cases shall be treated as Out-patients.

3. No person suffering from any Infectious Disease shall be admitted as an In-patient.

4. Friends of Patients are allowed to visit them on the following days:—

	P.M.	P.M.
WEDNESDAYS	6-30	to 7-30
SATURDAYS	2-0	to 4-0
SUNDAYS	2-0	to 4-0

5. No Patient is allowed to receive more than two visitors at the same time.

6. No smoking is allowed in the Hospital before 6 p.m., and only in the room provided for the purpose.

7. No stimulants, under any circumstances, are permitted to be brought into the Hospital by patients or visitors.

8. All articles brought into the Hospital by a Patient's friends or relations must be handed over to the Nurse in charge.

9. When in the Hospital, patients must behave in a quiet and orderly manner, and conform to the Rules of the Hospital. Any infringement of this Rule will render the patient liable to dismissal, at the discretion of the Resident Medical Officer.

By order of the the Water Committee.

E. ANTONY LEES,
Secretary.

Council House, Birmingham,
22nd February, 1895.

GEO. JONES & SON, TOWN HALL PRINTING OFFICES, 27-28, EDMUND STREET, BIRMINGHAM.

BIRMINGHAM CORPORATION WATERWORKS.

VACCINATION

NOTICE IS HEREBY GIVEN that the Corporation have made arrangements with DR. GORDON RICHARDSON, at the Accidents Hospital, in the Elan Village, to Vaccinate or Re-vaccinate, as the case may be (free of charge), all persons whose names are on the books of the Corporation.

All such persons who have been Vaccinated must call on the same day in the following week for inspection at the Accidents Hospital.

Days & Hours of attendance appointed for Vaccination:

Tuesdays and Fridays - 1 p.m. to 2 p.m.
Sundays - - - 10 a.m. to 12 noon.

By order of the Water Committee.

G. N. YOURDI,
RESIDENT ENGINEER.

March, 1896.

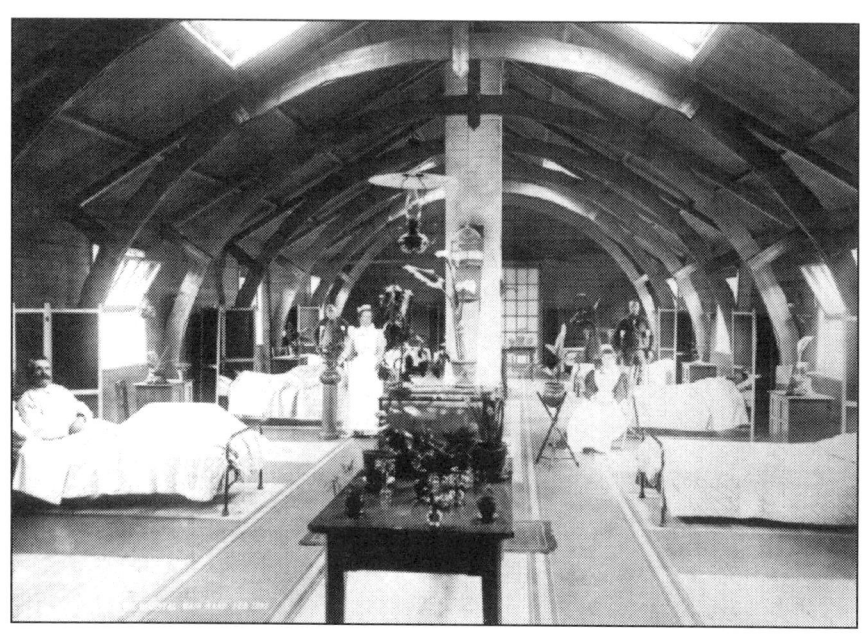

The Hospital wards

Inquests

Most of the accidents were caused by men falling onto rocks and by rockfalls. Those working in the stone mason's yard were susceptible to eye injuries.

A young boy was decapitated by a crane jib. The only comment made at the inquest was that he should not have been working. He had given his age as 14 when he was in fact only 11.

A young man aged 17 years from the village of Aba Cwmhir (which is near Rhayader) was killed by a train. His elder brother was also working on the train. At the inquest he stated that he had shouted to his brother and also sounded the whistle but he failed to hear the warnings.

Another inquest reported the death a young man who fell while pushing a wheelbarrow full of sand along a plank 18 inches wide and 20 feet up in the air. A verdict of sunstroke was given as the cause. The weather being very hot at the time. Eventually the verdict was altered to one of accidental death.

Cnwch Quarry. James Kingston was injured by an explosion. He died on May 6th 1896 from injuries caused by a charge of gunpowder exploding which he was putting into a hole.

John Evans, Labourer. Employed at the Crusher Yard and was killed while assisting another man to clear the shoot leading the crushed stone into the revolving screen. He was caught by the shaft and killed before the engine could be stopped. He had no dependants and no relatives.

Accident reported at Dlau Tunnel 2nd July 1896. Misfire charge went off. One man's skull is shattered and is not expected to live, two men will lose an eye, two slightly injured, one escaped with a shaking. Further report 7th July 1896, one man has died.

March 1901. Edward Gardner. Labourer. Employed in the Crushing Yard. Killed through the side of the hopper and Crusher breaking away through an overload of stone. Gardner had no dependants. His brother lived in Bristol and claimed £10.0s.0d. for his funeral which was paid by the insurance company.

An Inquest Report on Samuel Davies held on December 16th 1902 is described in great detail and the discription of the death of this man would not be suitable for public reading.

There is no true record of how many navvies died, mainly because none of their real names were known, being known mostly by nicknames.

The Canteen
The opening hours of the canteen were:

12.00 – 2.00
5.30 – 9.00 p.m. (weekday)
1.00 – 4.30 ⎫
5.30 – 9.30 ⎬ Saturday

RULES FOR THE CANTEEN-KEEPER

I. The Canteen-keeper must see that copies of the "Rules for the Management of the Canteen" are hung up in conspicuous places in the Canteen.

II. A return of all sales each day must be sent to the Chief Office before 12 o'clock on the following day, on a form provided for the purpose, accompanied by all moneys received.

III. Stock must be taken every Thursday, with the assistance of the duly-appointed ganger for the time being, and a return made of the same to the Chief Office.

IV. The premises must be cleaned out daily, and the windows cleaned once a week.

V. The Canteen-keeper is, under no circumstances, to drink with customers.

VI. Only the Canteen-keeper himself, and his authorised assistants, if any, are permitted to serve in the Canteen.

VII. The above Rules may at any time be added to or varied by the Water Committee.

By order of the Water Committee,
G. N. YOURDI,
Resident Engineer

Stock was taken every week, and the canteen keeper had to account, either in cash or in stock, for goods supplied and brought forward. The goods sold were limited to beer and porter (on draught and in bottles), aerated waters, tobacco and cigars. He had to guard against incivility towards customers, on the part of himself or his assistants, lack of cleanliness in the house and drinking vessels, adulteration of the liquors, selling out of hours and disorder and drunkenness on the part of the customers. If he was able to avoid offence in any of these respects, he was thought of no worse, if the takings fell off, and no better if they increased.

Before work commenced in August of 1892, a licence was granted to a farmer by Radnorshire Magistrates for the establishment of a public house to be called The Elan Valley Hotel. A footbridge was built across the River by the owner, the distance from the suspension bridge being one mile and 1,500 yards from the footbridge. The canteen was opened in September 1894, although the work was not fully complete.

The costs of the fittings for the canteen were:

Bar Fittings etc.	£ 95 13s.1d.
Self Registering Till	£ 33 5s.0d.
Clock	£ 4 15s.0d.
Tumblers, glasses etc.	£ 15 0s.1d.
	£148 19s.2d.

The takings, at the beginning of 1896 were between £41-£126 per week. However, in March 1898, two years after it had opened, the takings had accumulated to £14,750. The purchases for the same period being £9,250, making a profit of £5,500. The working expenses amounted to £1,932, cost of the building £720, plus stock amounting to £150. The nett profit therefore in three and a half years was 93%. The school received £730, and the expenses connected with Public Rooms and staff was £1,510, which included covering the salary of the Missioner. The canteen was a municipal public house, the only one of its kind in the country. The scheme of selling the beer to the Navvy was unique. In July 1896 a report stated that over a thousand gallons of beer was being sold per week or 20 hogs head.

The Manager had no interest whatever in the sale of drink. His salary was fixed, and he lived on the premises rent free. Out of his wages he had to pay his own barman and also for the cleaning of the house and vessels etc. All the goods for sale were ordered by the Secretary of the Water Department, on request from the canteen keeper, who did not know the real cost, as all the goods were charged to him at selling prices. There was an extension to the cellar in the canteen and samples were sent to Birmingham for analysis by a Professor Faulkner of Broad Street for which he received ten guineas per year. The cases to be tested were marked with numbers so that no-one knew which of the brewers beer was purchased. On 5th October 1896 a letter was written by Mr. Edmund Gwynne, Chief Constable of Breconshire, praising the committee regarding the canteen.

The canteen was situated almost in the centre of the village entrance being restricted to men. There was plenty of drinking done and it was not unusual for several navvies to be thrown out each night for drunkenness. Not to be outdone, the navvies built a small bridge across the River Elam, so it was out of the canteen across the river to Rhayader where they got truly drunk making it impossible for them to get home. Most of them spent the night sleeping in the hedgerows, which was fine during the summer but many of them died from hypothermia during winter months.

CITY OF BIRMINGHAM WATER DEPARTMENT.
ELAN SUPPLY.
ELAN VALLEY WORKS.

RULES FOR KEEPER OF THE CANTEEN
IN THE ELAN VILLAGE.

No. **1** Hours open.	(*a*) The Canteen will be opened every working-day in the week (Saturdays excepted) between the hours of 12 noon and 2 p.m., for one and a half hours only; and for the whole time in the evening between 5-30 and 9 o'clock.
Saturdays only.	(*b*) On Saturdays, the house will be opened from 1 to 4-30 and from 5-30 to 9 p.m.
Hours closed.	(*c*) The house will be closed every evening at 9 o'clock prompt.
Sundays.	(*d*) On Sundays the house will be closed all day.
No. **2** No women allowed in the bar.	Women will not be allowed in the bar at any time, under any pretext whatever.
No. **3** No boys in bar.	Men only over 18 years of age will be permitted in the bar.
No. **4** No women under 21 to fetch drink from Jug Department.	(*a*) No woman under the age of 21 years will be served with beer or porter at the jug department.
No boys under 16.	(*b*) No boy under the age of 16 years will be served with beer or porter at the jug department.
No. **5** No drinking in Jug Department.	No one will be allowed to remain in the jug department beyond the time necessary to obtain the liquor required, and no one will be allowed to drink in that department.
No. **6**	No person shall be supplied with more than one quart of liquor at the morning hour.
No. **7**	No person shall be allowed more than two quarts of liquor in the evening for consumption on the premises.
No. **8**	No person who is in the slightest intoxicated shall be supplied with drink on any pretence whatever.
No. **9**	All persons applying for liquor at the jug department must be duly registered inhabitants of the village, and no person shall be supplied who is not such.
No. **10**	No hut-keeper shall be supplied with more than 1½ gallons of beer in any one evening, nor with more than two gallons for the mid-day meal from the jug department, except on Saturday evening, when a hut-keeper may purchase double the quantity.
No. **11**	Only village residents shall be served at the bar, except on a written order signed by the Resident Engineer.
No. **12** No amusements in the house	Amusements in the house are strictly prohibited. No music, singing, juggling, reciting, gambling, card playing, playing dice, dominoes, draughts, marbles, shovel-penny, or any game either of skill or chance, will be permitted in the house.
No. **13** Disturbances.	In case of any disturbance or quarrel in the house, the parties will be immediately ejected. The Canteen-keeper and Village Superintendent have strict orders to turn out any disorderly person, and to close the house if order cannot be maintained.
No. **14**	The above rules may at any time be added to or varied by the Water Committee.

BY ORDER OF THE WATER COMMITTEE.

G. N. YOURDI,
RESIDENT ENGINEER.

Outside the Canteen

Letter dated January 18th 1894
Received by the Water Committee from Mr. Yourdi

Shelter sheds upon the works are, I need not point out, greatly appreciated by the men especially when fitted with a flat-topped stove where food and drink they bring with them can be warmed.

A shed partaking central character is not to be recommended but a series dotted here and there at points easily reached by the men in case of heavy showers. The sale of coffee and soup would be best left to private enterprise and facilities given to the interested vendors of these commodities who as a rule follow in the wake of navvies. In the summer time the Committee will of course follow the customary practice of giving the men free of charge oatmeal and water. This course is generally adopted by humane and far-seeing contractors for the purpose of not only giving the men an opportunity of slaking their thirst but preventing the men leaving their work for short intervals on the plea of getting a drink at the nearest spring.

I would therefore advise a series of cheap shelters in the neighbourhood of the dam and fitted on the lines suggested where coffee and soup might be sold by the vendors who are sure to come up once it is know that facilities will be given them to carry out their business at certain hours in the day.

Temperance Hotel, Rhayader
December 4th, 1893

Dear Sir

Excuse me taking the liberty of writing to you but the report of Alderman Danby Parker in the Daily Post of November 30th seems to have ?????? tradespeople in Rhayader not erecting shops at the village at Caban Coch but leaving it to them to supply the demand. I am rather afraid the shopkeepers will charge rather heavy prices for their goods unless there comes some competition from outside towns then they may do it as there seems to be an idea amongst them that they are all going to make a large fortune out of the working men that are coming here to work in fact already most of the owners of houses and shops where have their Tenant's Notice for a rise in their rents and in tone case to pay the last half year advance or leave but no doubt there will be outside competition which will probably keep them without charging too high a price for their goods. I was also pleased to see the committee had decided to open a beer shop at the works as no doubt it will prevent a lot of secret selling of beer up the valley but probably it may be best not to sell too strong beer up there but a light beer that would be best for the men while at work. I hear private selling is going on already near Aber Caethon. The Coffee House no doubt will be very beneficial to the men but I think too that if the committee was to put up a dining room in connection with it to hold a hundred or perhaps two hundred where the men could get a cheap cooked dinner or a basin of soup would be a step in the direction of benefit to the men and would be able to work better that with only the dry food they carry with them the room also may be used in the evening for reading rooms or for various amusements that may be provided or for mission room on sundays or various things if properly carried out may be of great benefit to the men as no doubt many may be employed there that may not live in the village and a suitable place for these to get their meals in the days also for the benefit and enjoyment of those that live in the village in the evenings.

I am only stating my opinion privately to you but if you think the idea may be of any service to the committee you may mention it to them as from what I can see the committee are doing what they think best for the benefit of the men.

Yours faithfully
J Hartland

Shops in the Valley

The Water Committee were approached by the residents of the Elan village for a Co-operative store to be opened but on advice from Mr Yourdi this was rejected.

While preparations were being made for this marvellous feat of engineering, a young man, Richard Hughes, whose ancestors had been shepherds in the valley, arrived in Rhayader to open a draper's shop. He rented premises from Mr Lewis Lloyd, a wealthy landowner who, with a Rev. Prickhard, owned a large area of the Elan Valley. Mr. Hughes was granted permission by Birmingham Corporation to open a shop in the village.

R. HUGHES.
GENERAL DRAPER & GROCER
BOOT & SHOE WAREHOUSE

London House

Rhayader Dec 10 1895

F. E. A. Dees, Esq.

Sir,

I am about to build a place for business on land belonging to R. Lewis Lloyd Esq. I should feel obliged to the Water Committee if they would allow me to pass over the bridge and as my shop adjoins Mr. R. Worthing's near the Bath House (West End) and I am informed he has made application to you for an entrance from the Village to his proposed shop. I should feel grateful to the Committee if they will allow me to use the gate which he has applied for. I would pay the Corporation and acknowledgment for so doing. My term of tenancy would be but for a few years. I would comply with any regulations you wish carried out, also allow your Inspector to visit my place of business and report on the land.

I may now say I hold a permit to enter the village.

I am, yours respectfully,

R. Hughes

Old shop in Rhayader

New shop in Rhayader

George Griffiths, manager of shop in Elan Village

As the work progressed in the valley Mr Hughes put in another application to open a second shop at Pen-y-Garreg. This application was again successful.

Old shop in Elan Village

Shop in Pen-y-Garreg before conversion

Shop in Pen-y-Garreg

With the completion of the scheme the dams began to fill with water. The drapery shop at Pen-y-Garreg began to float whereupon Mr. Hughes rescued it and rebuilt it on the side of the dam. This building was later used as a Mission Hall and remained so until 1965 when it was demolished, the seats being removed to the Church of St. Bride's where they remain to this day. The family drapery business continues to flourish in Rhayader and is managed by grandson, Richard.

A general stores was opened in the village soon after Mr Hughes opened his shop. A report shows that a complaint was lodged that short measure was being given in the sale of milk and butter. Very strict regulations were in force with regard to the construction of the stores. Oil for lamps etc. was not allowed to be sold and storage of it for use within the shop was very restricted, the standard of hygiene being very high. The water supply was taken from the Corporation's mains at the expense of the applicant.

Church

In November 1895 there was great concern by the church about the spiritual needs of employees of the Water Committee and their dependants. The church of Nant Gwillt would eventually be submerged when the valley was flooded, but until such time the clergy requested that there should be an extension to the church to accommodate the increasing population.

There was a Gospel Mission Room built on ground adjoining the Elan village in the year 1895.

On the downstream side of Garreg-Ddu was the Baptist Chapel which was used as a Social Institute for staff. It was replaced by a new building below the Caban Dam which was erected at a cost of £1,127. 10s. 2d.

Before the church of Nant Gwillt was submerged records show that five bodies were removed, all relatives of a Mr. Lloyd. The removal of the remains from the old graveyard and re-interment at the new site was carried out by the Corporation at their expense. The conveyance of the bodies, however, together with the cost of the coffins, was paid for by Mr. Lloyd. An act of 1892 stated that a sum of not more than £10 was to be paid by the Corporation for each body.

CHAPTER THREE

The Engineers

JAMES MANSERGH, F.R.S.

James Mansergh was a Past President of the Institute of Civil Engineers and also a Justice of the Peace for the county of Radnor. He was High Sheriff for that county in 1901.

Born in Lancaster on April 29th, 1834, he was educated at Harmony Hall, Hampshire. His first leaning towards a career in engineering was grained from one of the masters as his school, John Tyndall. James was a broad-shouldered distinctive man with a fine face, flowing beard and massive brow, a man of commanding presence and cut quite a dash in knickerbocker suit and an alpenstock in his hand.

He was apprenticed in 1849 to Messrs. H. McKie & Lawson, engineers and surveyors, of Lancaster, where he gained a wealth of experience which stood him in good stead for his chosen career. At the age of 21 he went to Brazil as engineer to a Mr Price and worked on the Rio de Janeiro railway.

In 1859 he returned to England and joined the company of McKie at Carlisle engaging in general engineering and designing the first sewage farm in England. In 1862 he returned to railway work as a Contractor's Agent for Messrs. John Watson & Co. his first project being the mid-Wales and the Llandis and Carmarthen railway carrying out the duties of railway engineer.

Early in 1866 he entered into a partnership with his brother-in-law, John Lawson, of Westminster. His first task was designing a gravitation scheme to supply water for Carlisle. Mr Lawson advised Birmingham upon a water supply for the city. Mr Mansergh, some 20 years later, suggested a similar scheme utilising the valleys of Elan and Claerwen with which he had become acquainted while working on the mid-Wales railway eight years earlier.

After the death of his partner in 1873 Mr Mansergh designed and constructed sewage works throughout the country and was in constant demand in connection with Parliamentary work. He was also called upon to carry out work in many countries throughout the world and at home, designing sewage works and water treatment plants and also giving advice.

He acted for no fewer than 360 local bodies and prepared over 250 reports. He gave evidence at some 300 public inquiries and attended upwards of 600 appearances at both Houses of Parliament.

James Mansergh, F.R.S.

It is in connection with the Elan Valley water scheme, however, that Mr Mansergh will best be remembered. A visit to the Elan Valley gives an insight into his genius. Mr Mansergh was too occupied with professional work to attempt much writing but he did give the occasional lecture.

In March 1903 he was made an honorary freeman of his native town of Lancaster. In acknowledging the honour he said: "I cannot take to myself all the credit that has been given me. My success has been mainly due to the opportunities I have had. I was brought up in a God-fearing household and its influence upon my life has never been lost.

The only credit due to me is that I have worked hard and steadily and at the same time been honest and straightforward.

Mr Mansergh died on June 15th 1905 at his home in Hampstead, London.

GEORGE YOURDI

George Yourdi was resident engineer for the building of the Elan Valley dams and responsible for all work upon the water shed.

He was not very tall and had a dark complexion with a clean-shaven face save for a heavy black moustache, silver-streaked black hair and flashing eyes. He had a penetrating keenness for spotting careless workmanship by his staff.

Mr Yourdi had an appearance which at once impressed all who met him. He was of Greek extraction. His father was a Greek consul at Cork and his mother was Irish, so he possessed good looks and some of the racial characteristics of two highly interesting and imaginative nations. A Fellow of Trinity College, Dublin, a Bachelor of Arts and Bachelor of Engineering, Mr. Yourdi worked strenuously at his chosen profession of which he was without doubt an expert. He had a considerable amount of experience on water undertakings being an expert in cement work and an authority on matters relating to their constituent parts. This, of course, was particularly important with regard to dam building. He was a bachelor and so were most members of his staff.

He threw himself wholeheartedly into his work. Indeed for eight years it is said he never spent a single night out of the valley. Every detail of this huge project had been under his control.

Mr Yourdi tramped up and down the valley for miles inspecting, directing and controlling his staff during the coldest days of winter and the hottest days of summer.

George Yourdi

CHAPTER FOUR

Characters of the Valley

MR. H. ATKINSON, ENGINEER OF GARREG DDU

Mr. Herbert Jefcoate Atkinson   Miss Mary Kathleen Jane Ashe

HERBERT'S father was a linguist in the second half of the nineteenth century and could speak about twenty different languages and taught among many others, students and staff for the Colonial Office. He was a Professor of Sauskist and Modern Romance languages at Trinity College, Dublin and had previously been a student in Paris.

Mary's father was a Surgeon whose name was Isaac Ashe who also worked in Dublin. When Herbert asked if he could marry Mary, her father replied Yes! but only when you have a respectable job.

After waiting seven years he was offered a position by Mr. Yourdi who also came from Ireland.

Herbert and Mary married and went to live in Elan Valley where they had four children. A fifth child was born after they left the valley.

The two boys and three girls were all given an extra surname and the second daughter being named Margaret Yourdi Atkinson as Mr. Yourdi was her god-father.

The children were educated at home by their mother until the age of about 10.

When the Dams were completed the family moved to Manchester where Mr. Atkinson took up a position of City Water Engineer to the Corporation.

Mrs. Atkinson died in 1916 as a result of pneumonia, and the eldest daughter took over her mother's position and missed eighteen months of school. Her younger brother and sister were still at home when she went to university in 1918. The family had by then acquired a housekeeper to help out in the household arrangements. After completing her studies she qualified as a doctor – giving up her profession on her marriage and moved to Birmingham.

Mr Atkinson

Site of original Elan Village, Garreg-Ddu (note old Baptist Church in left foreground)

The Chitty Family History

A solicitor's clerk living in Birmingham applied for a position of wages clerk on the works in the Elan Valley. At the time of his appointment there was no accommodation available (the village had yet to be built). He was fortunate to find rented accommodation in Rhayader for himself, his wife and five children. There was no public transport available in those days; he walked to the valley and back each day in all types of weather. Eventually he was offered a house in the village, No. 1, and settled there with his wife and family which by then had been increased by a further two children.

Life was hard for the womenfolk in those days. There was no hot water, microwaves, washing or drying machines or electric appliances. When the eighth child arrived the poor mother was almost suicidal, because there was no room for the new baby so its cradle was suspended from the ceiling and brought down by a pulley when it was time for it to be fed and changed and then raised again by pulley. When the baby cried the cradle was rocked with a broom handle. Apart from looking after the family, Mrs Chitty also baked bread for the villagers. At the end of the day when everyone else was in bed her treat was to put her feet in the oven to get them warm. Her husband also took on responsibility of looking after the canteen which was the modern-day equivalent of a public house. It had a stone floor like all the other buildings in the village, and this was then covered with sawdust. Albert, the third youngest, was given the job of raking the sawdust every morning and allowed to keep any money he found.

Mr. Chitty (left) and Mike Bennett, Manager of Elan Works

Eventually the dams were completed and the family remained in the village, with the head of the house taking on yet another job as rain gauge keeper. The wooden huts in the village were replaced with beautiful stone-built houses you see today and the family moved into a double-fronted one. This served as a shop and also provided bed and breakfast accommodation.

Accommodation at Pen-y-Garreg

As the work progressed accommodation was badly needed. Cwm Elan house was over-crowded. There were nine engineers living there when when there was only room for eight.

A building was erected at a cost of £750 for seven engineering staff engaged in the construction of the Pen-y-Garreg Dam and Craig Goch. It was to be furnished and inhabitants were charged a rent leaving them to provide their own board and service.

The building was built between Cwm Elan and Pen-y-Garreg, thus making it central for the engineers engaged at both dams.

The cabinet furniture was supplied by Messrs. John Haugh & Son, Messrs. Hollidays for upholstery, bed and table linen and cutlery. Messrs. Hollidays giving a special discount.

At a later date a request was made for a laundry room to be built. Permission was given. It cost £70 and this was met by the Water Committee. The engineers had to supply the soap.

Pen-y-Garreg House, home of Mr. Tickell and a number of other engineers.

Mr. Eustace Tickell

Very little is known of Mr E. Tickell and yet what is known is very interesting.

He was born on 29th April 1864 and at the age of 20 studied at the technical school in Hanover. In 1887 he went as a pupil under Mr Mansergh and gained experience in the water industry under the supervision of Mr Yourdi who was himself employed by Mr Mansergh.

In 1893 Mr Tickell was sent with Mr Mansergh's two sons to survey the Elan-Claerwen Valleys. He took with him a sketch pad, pen and ink and sketched the valley before construction work began. He also wrote a book to accompany the sketches which were part of a presentation book and given by the Water Committee to those landowners who had co-operated with Birmingham in allowing the pipeline to be carried through their estates. These are now collectors' items.

Soon after the work commenced in the valley Mr Tickell was appointed engineer in charge of the Pen-y-Garreg Dam receiving a salary of £250 per annum, later increased to £300. He resided at Pen-y-Garreg House with seven other engineers having to pay rent and the wages of servants (7s. 6d. each per week).

When the dams were completed Mr Tickell then went to carry out work on building railways in Ceylon.

He died on 8th September 1943.

Mr. Tickell, Engineer in Charge of building the Pen-y-Garreg Dam

Above and on page 68 are two of the wonderful sketches of the Valley produced by Mr. Eustace Tickell

Mr Lloyd, Resident Engineer of Craig Goch

CHAPTER FIVE

The Building of the Dams

WAGES

In October 1895 it was stated that a Clerk was sent from Birmingham once a fortnight by the mid-day train on a Thursday with the men's wages and by arrangement with the Bank Manager the cash was counted on that same evening and locked in the corporation cash bag, the clerk retaining the key. The bag was then locked in the safe. On Friday morning the Clerk then counted the money for each man. With the increase of numbers of men to be paid it was decided by the Committee to pay the Bank Manager, Mr. Rigby £50.00 per year which would be shared amongst his clerks and it was agreed that the Clerk should be sent on a Wednesday night enabling him to commence work at 3.00 p.m. on Thursday with the help of the Bank Manager and his staff.

Wages for some of the work-force

Engineer Mr. E. Tickell	£250.00 per annum	
Assistant Engineer Mr. E. Fenby	£13.00 per month	
Junior Clerks Cost Department	25s. per week	} June 1895
Boy Clerk Cost Department	14s. per week	
Chief Cost Clerk	£130.00 per annum	} 1895
Missioner	£150.00 per annum	
Stonemason	7d.-10d. per hour	
Carpenter	5d. per hour	
Labourer	4d. per hour	
Boy	2d.-3d. per hour	

It was reported that five men worked one hour from 5.30 a.m. until 6.30 a.m. boiling water. A boy was employed for ten hours carrying tools.

Reports of other payments made around the Village

Caretaker at Hospital – E. Allen   January 6th 1903 to February 3rd 1903. Four weeks at 21s. less four weeks board at 15s.   £3.00   Pay £1.4s.0d.

Blacksmith – Frank Morgan
Frost Nailing Horse 1s.   Fastening two shoes 6d.   Total 1s.6d.

Chimney Sweep – Cleaned two chimneys at Doss House 6d. each – 1s.

Pay Mrs. Thomas £1.10s.0d. for assisting at the funeral from Cwm Elan. From 10th March 1904 to 23rd March 1904. Less 1s. per day for food.

```
     Birmingham Corporation Water Works.   A 209
     REQUISITION FOR PREPAYMENT OF WAGES.  £ 4

  Pay No.              Date


  Timekeeper
                  Cashier
```

STONEMASON

The stonemason was the highest paid worker apart from the engineer engaged on the construction of the dams. For all his hard work he received between 7d. and 10d. per hour. Most of his work is under water.

Stonemason's work

The stone was blasted away with dynamite and then brought to the stonemason's yard. Some of the men worked by standing on plat-forms surrounding large pieces of stone while others worked in open sheds. A crane would bring the pieces of stone for cutting and a door in the roof of the shed would be opened and the stone placed in front of the mason who would then cut the stone to the measurements given.

GENERAL WORK

The Carpenter was paid 5d. per hour. He played a very important part in the construction of the Dams. The Labourer was paid between 3d. and 4d. per hour. The hours worked were fifty to sixty hours per week depending on the time of the year. Should the weather prevent work being done on the dams construction the men would be put to building roads.

Mr Yourdi was a meticulous man and so were the inspectors which meant that all work had to be of a very high standard. The concrete had to be of a certain mix. All bricks an exact size and the cement between a certain thickness. If these details did not meet with the set requirements the work had to be done again. All men engaged on the construction of the dams, whether they were labourer or engineer, took pride in what they did giving of their highest standard. Their work is in the valley for all to see.

The Stonemason's Yard

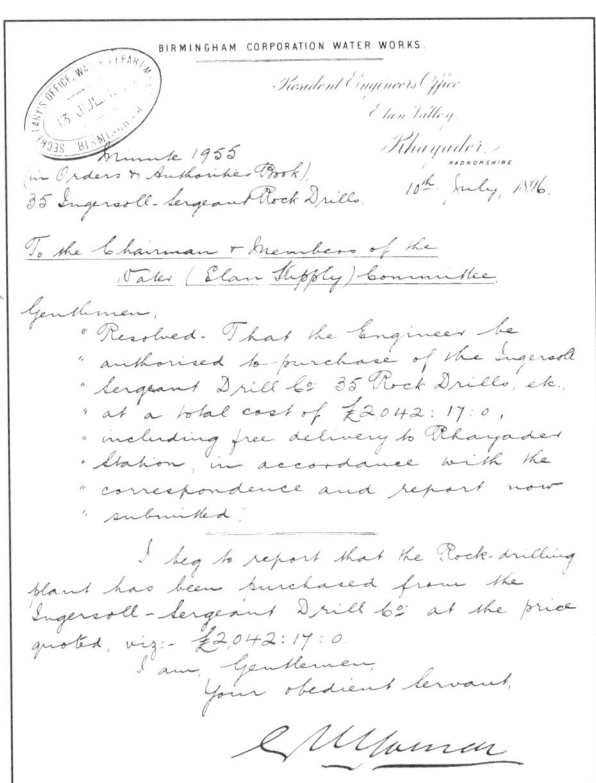

BIRMINGHAM CORPORATION WATER WORKS.

Resident Engineer's Office
Elan Valley.

Rhayader
RADNORSHIRE.

Minute No 5385
(In Orders & Authorities Book.)
 Blue Pressed
 Invert Blocks

3 June 1901

To The Chairman and Members of
the Water (Elan Supply) Committee

Gentlemen,

Resolved "That the Engineer be
authorised to purchase of the Hamblet's
Blue Brick Co Ld. West Bromwich, 34,000
Blue Pressed Invert Blocks at a cost
of £20.6.0 per 1000, nett, delivered Rhayader
Station."

I beg to report that 34,733
Invert Blocks were purchased at a
total cost of ~~£704.1.7~~ £705.0.9.
(corrected as per Mr Yourdi's
letter of June 6/01)

I am, Gentlemen
Your obedient Servant

J M Yourdi

Blue Bricks Lining Cut & Cover

BIRMINGHAM CORPORATION WATER WORKS.

Resident Engineers Office
Elan Valley.
Rhayader
RADNORSHIRE.
15th July, 1896.

Minute 793
(in Orders & Authorities Book).
Cement Cooling Shed.

To the Chairman & Members of the
  Water (Elan Supply) Committee.

Gentlemen,

"Resolved. That the Engineer be authorised
to erect a Cement-cooling Shed, in
accordance with the plan now submitted,
at an estimated cost of £2,300."

    I beg to report that the Cement
Cooling Shed has been erected at a cost
of £2,428:17:2, inclusive of platforms and all internal
fittings excepting Easton Anderson & Goolden's
a/c for hydraulic jiggers and the cost of
erecting same, being £128:17:2 in excess
of the estimate.

    I am, Gentlemen,
        Your obedient Servant,

The Cement Shed

Steam Crane

STEAM COAL

Name	Description of Coal	Price previously supplied at	Price now quoted
Crawshay Bros.	S. Wales	16s. 3d.	16s. 3d.
Old Radnor Co.	Best Dowlais	15s. 9d.	15s. 3d.
W. Burgum	Cannock	15s. 6d.	14s. 2d.
Radnorshire Co.	Black Fark – North Wales	14s. 2d.	13s. 1d.
Radnorshire Co.	Cannock	15s. 6d.	15s. 0d.
Wm. Morris	Shropshire – S.W.	15s. 0d.	15s. 0d.

The coals which have given most satisfaction are those supplied by Crawshay Bros. and the Old Radnor Co. – South Wales Coal – the North Wales Coal being unsatisfactory in all cases. The Cannock coal supplied by Wm. Burgum was also unsatisfactory.

SMITHY COAL

With regard to the Smithy Coal none has yet as been found satisfactory, though coal has been obtained from the following firms:

Name	Description of Coal	Price previously supplied at	Price now quoted
Crawshay Bros.	South Wales	15s. 6d.	–
G. J. Stothert & Co.	South Wales	13s. 3d.	12s. 9d.
Wm. Morris	North Wales	13s. 3d.	–
Old Radnor Co.		15s. 3d.	–

A copy of a letter about coal supplies

Steam Crane

Report given by Mr. Mansergh's sons
Ernest Lawson Mansergh and Walter Leahy Mansergh

The Dams were built of cyclopean rubble faced with rock-faced block-in-course masonry. The rubble which consisted of plums ranging in weight from 1 cwt to more than 10 tons were set in Portland cement concrete gauged generally 5-1. The stones in the facing range from 18ins. to 4ft. in width on the bed and were backed with 4-1 concrete for a thickness of 5ft. at the up-stream face and with 6-1 concrete 3ft. thick at the down-stream face. The concrete was all mixed by hand and of thoroughly clean materials and then well rammed between the plums until the whole was in a state of quiver. Plums were all washed and brushed perfectly clean, and their bases were roughly dressed to ensure level bedding.

Practically the whole of the footings for all the dams was local grit and conglomerate obtained from two quarries, one at each end of the Caban Dam, but some of the facing stone had to be imported from quarries near Pontypridd and Builth Wells.

The cement used was brought from the River Thames by barge to Aberdovey, and in the specification it stated all sand and every brick to be washed before being put into place. Maybe this is the secret of why the dams are still there in perfect order. The men did as they were told and took a pride in what they did.

Each month during the construction of the dams. Reports were sent from the Valley to Birmingham. One outstanding report stated. Progress on the tunnel difficult this month we have travelled one foot in 10 hours.

Many tickets were found for navvies' Monkey Jackets. They were white with the letters B.W.D. on the back of them. A man would exchange 2s. 6d. for a Jacket and a number, this being entered into the Record Book. When his contract finished the jacket was returned in exchange for the 2s. 6d. In later years these jackets took the name Donkey Jacket.

CHAPTER SIX

Expenses incurred during building

HUNDREDS of receipts were found in an old leather-bound book. A few samples provide both interesting and amusing reading.

Purchased from Thomas Evans, Rhayader
April 26th 1902 – 10 slates 18 x 10 – 2d. each 1s.8d. for Stable Cottage
April 30th 1902 – 18 slates 20 x 12 – 2½d. each 3s.9d. for Bethania Chapel

Thomas Warner – Bridge Street, Rhayader
   1 dozen egg cups at 1d. each – 1s.
   3 china cream jugs 6d. each – 1s.6d.
   3 tea pots at 9d. each – 2s.3d.

The Bridge, Elan Village to Mr. Jones at Cwm Elan Mines
   One fowl   2s.3d.
   1 dozen eggs 1s.
   25 eggs   1s.
                    4s.3d.

Price Stores, Elan Village, Near Rhayader, Grocery, Drapery and Provisions

Nurse Parkes – February 2nd 1904
One pair of pants	2s.11d.
One vest	2s.11d.
One cloth cap	1s. 0d.
One pair of bracers	1s. 0d.
One neck-chief	6d.
One pair of boots	8s.11d.
	17s.3d.

Edward Morgan – Dol-y-Mynach Isolation Hospital – January 1903
Mutton 13½ lbs at 9d. per lb – 10s.1½d.

Compensation paid
May 12th 1903 – one lamb and two sheep killed by train
August – one lamb killed by train
September – one sheep killed by train
October – three sheep killed by train
December – two sheep killed by train
                        £9.10s.0d. paid to Evan Jones

January 17th 1895 – bought for Mr. Brightmore's use
Horse   £35.0s. 0d.
Trap   £33.0s.11d.
Harness etc.   £21.7s. 5d.
               £89.8s. 4d.

Mr. Yourdi
March 1903 – Lion Hotel, Rhayader
Stabling for one year – £1.1s.0d.

January 15th 1903
1 hand lantern purchased for Filter Beds – 2s.7d. including postage.

Mr. Yourdi
Reading Glasses 3½" dia. at 7s.0d. each   14s. 0d.
                         Discount 10%   1s. 5d.
                                             12s. 7d.
                            Postage   4d.
                                             12s.11d.

Reports
December 1903 – Smallpox outbreak

   Mrs. Brown and her daughter from Hut 22, Pen-y-Garreg washed the whole of the bedclothes of the Corporation and J. D. Hughes whilst Hughes and his family were in quarantine for which she charged 12s.0d. This a reasonable in my opinion.

                            Signed Yourdi

   I beg to report having spent the following sums of money in connection with the deaths of the following persons:

   John Jones died on January 24th 1902 from falling from gantry at Caban Dam.   Conveying body to Mortuary   3s.0d.
              Telegram and reply to father   1s.9½d.
                                                  4s.9½d.

   John Ashton – November 19th – Telegram and reply.
This man was buried by Parish having only worked 10 days – 1s.7d.

   Samuel Davies – December 13th – Quarry Explosion.
Telegram and reply   1s.4d.

   On the recommendation of Mr. Mansergh, the Water Committee appointed the firm of Hudsons of White Chapel, London to take photographs of the dams during the construction. The payment being £1. 30s. 0d. for every hundred photographs taken.

CHAPTER SEVEN

The Opening

HIS Majesty King Edward VII and Queen Alexandra visited the Elan Valley on July 21st 1904 for the opening of the new water supply for Birmingham. The ceremony was attended by 750 proud and honoured guests.

Three trains (red, white and blue) left Snow Hill for Rhayader, the journey taking about three and a half hours.

The red and white trains arrived on time but the blue one had to be shunted into a siding at Rhayader to allow the royal train to pull in, it being late arriving from Swansea, the king and queen having spent the previous day there.

The belated blue train and its passengers did, however, manage to arrive at the sand filter beds to see the actual turning on of the water after which the King knighted the Lord Mayor. One could feel that the wrong person received the honour. Mr Mansergh, the designer of the whole project, who had made it all possible, received nothing but a thank you.

From the filter beds the royal train took their majesties past the dams, Caban Goch, Garreg-Ddu, Pen-y-Garreg and Craig Goch. They then returned to Nant Madog Field for lunch, which consisted of the following:

Croutons of Caviare
Lamb (hot) Cutlets
Sauté Potatoes
Medallions of Quails à la Royale
Cold Lamb
Roast Chicken, York Ham, Tongue
Galantine of Chicken Truffled
Pressed Beef
Potatoes, Salads
Strawberry Cream Helene, Ices
Dessert
Coffee

While their majesties and guests were enjoying the meal they were entertained by a Welsh choir from Llandrindod Wells and a band of harps.

City of Birmingham Water Department.

VISIT OF THEIR MAJESTIES THE KING AND QUEEN
TO THE ELAN VALLEY WORKS, RHAYADER,
AND
INAUGURATION OF THE NEW SUPPLY BY HIS MAJESTY THE KING,
JULY 21ST, 1904.

OUTLINE OF ARRANGEMENTS.

The party from Birmingham will be conveyed by special trains as under:—
From New Street by L. & N. W. R. at 6-20 and at 6-40 a.m.
"   Snow Hill (Livery Street Side) by G. W. R. at 7-25 a.m.

The journey to Rhayader will occupy approximately 3½ hours. The allotment of the visitors to particular trains will be made by the Water Committee after the number of acceptances has been ascertained.

Breakfast will be provided on the L. & N. W. R. trains at a charge of 2/- each.

The visitors will be met at Rhayader Station by trains to convey them over the Corporation Railway to the Foel Filter Beds where the Inaugural Ceremony will be performed.

The Royal Party will arrive at the Filter Beds at 12-30, and immediately after the ceremony will proceed by the Corporation Railway to an inspection of the Reservoir Works.

During the absence of the Royal Party on their inspection, the other visitors will be conducted to the general luncheon marquee.

On their return the Royal Party will proceed to the special Pavilion provided for their use, where they will take luncheon on the invitation of the Lord Mayor, and at the same time luncheon will be served in the general marquee.

The Royal Party will leave for Rhayader at 2-50.

After the Royal train has left, the visitors will be conveyed in Corporation trains over the Royal Inspection Route. Twenty minutes will be allowed at Craig Goch to leave the trains and inspect the Reservoir and Dam.

At the conclusion of the tour the Corporation trains will proceed direct to Rhayader where the visitors will take their places in the same trains by which they travelled in the morning, and will at once depart for Birmingham.

A meat tea will be served on the trains during the return journey.

The L. & N. W. R. Company's trains are timed to reach Birmingham at 8-10 and 8-45 p.m. respectively, and the G. W. R. Company's train at 9-26 p.m.

Further particulars, with programme of the proceedings, will be forwarded in due course to those who accept the invitation.

E. ANTONY LEES,
Secretary.

44, BROAD STREET,
BIRMINGHAM,
30th June, 1904.

The Royal Train arriving at Filter Beds

*Their Majesties turning on the water.
Mr. Mansergh, King Edward, Queen Alexandra and The Lord Mayor*

The Banqueting Marquee

The King's Changing Room

The Queen's Changing Room

CHAPTER EIGHT

What is water?

CHEMICALLY, water is hydrogen monoxide with the symbol H_2O and is a combination of hydrogen and oxygen, both of which are gases. Oxygen is almost 16 times as heavy as hydrogen and the two gases combine to form water in the proportions by volume of one volume of the former to two of the latter. Thus the proportions by weight in which these gases exist in water must be about 16 to 2. Most careful experiments have shown that 88.86 parts of oxygen by weight unite with 11.14 parts of hydrogen by weight to form 100 parts of water.

Water is an extraordinarily good solvent and there are but few products of nature which can resist its continuous chemical and mechanical action. It is due to its capacity for absorption that water is never found absolutely pure in nature, foreign matter being either gases, minerals, organic or mechanical mixtures, one or other predominating according to the locality in which the water is found.

In a state of absolute purity, water is a clear, colourless, transparent and odourless liquid, which undergoes surprising transformations at different degrees of heat and cold. At a temperature of 100°C. (212°F.) it evaporates in steam, while at a temperature of 0°C. (32°F.) it freezes into a hard, solid crystal mass. Distilled water is 815 times as heavy as air and its density is represented by unity, or 1.0, which is taken as the standard of comparison for all liquids and solids, just as hydrogen gas serves as the standard for gases and vapours.

THE ELAN AQUEDUCT

The Elan Aqueduct passed through five counties: Radnorshire, Herefordshire, Shropshire, Worcestershire and Staffordshire.

Birmingham receives 80 million gallons of water from the valley every day and there is no pumping whatsoever. It is all done by gravity feed.

The fall from the valley to Frankley is only 171 feet. The water is carried by four pipes two of which are 60 inches in diameter and the other two 42 inches. It also travels through a tunnel and what is known as cut and cover. The journey takes two-and-a half days and the water never sees daylight from the time it leaves the valley until it arrives at the Frankley Reservoir.

The Treatment Works at the valley were not in the original design but it was found that the peat content in the water would corrode the cast iron pipes.

It was decided to remove the peat by adding lime to the water. The next stage of the treatment was very simple. The water is passed through sand filters, the sand working like a tea strainer by collecting wool, leaves etc. Fluoride is added at the next stage as a benefit for children's teeth. This is done at the request of the Birmingham Health Authority. The last stage is the addition of chlorine to kill germs.

On arrival at Frankley the water is put through sand filters. No lime or fluoride, only chlorine is added before it goes on to distribution which again is by gravity feed. Only two areas receive their water by pumping Northfield and Warley.

Conduits

The conduits are designed to operate with "free water surface", i.e. when delivering water at their maximum capacity the water level does not reach the roof. They comprise two types of construction, tunnels and "cut and cover", the finished structure of both being much the same, but the latter being built in open cut, subsequently refilled. There are fifteen tunnels of horseshoe shape, the two longest being $4^1/_2$ miles and $2^1/_2$ miles respectively, and the total length approximately 12 miles. The internal dimensions are approximately 8 feet high by 7 feet 6 inches wide. The gradient of the long tunnels is 1 in 3,000 and of the remainder 1 in 4,000. The sixteen lengths of "cut and cover" are similar in shape and dimensions to the tunnels.

The conduits are lined throughout with concrete, faced internally on the sides and invert with blue brickwork, and have a maximum designed capacity of 75 masonry piers and at two points where the conduit had to be carried over roads the necessary headroom could only be obtained by laying steel barrels 8 feet 6 inches in diameter between masonry abutments.

Siphons

The siphons, 11 in number, are constructed as pipelines across the larger valleys, and longest being $17^1/_2$ miles in length. The gradient is generally 1 in 1,760 or 3 feet per mile.

The original scheme envisaged six parallel lines of mains and as a first instalment two of 42 inch diameter were laid to convey 25 million gallons a day. They are of cast iron pipes where they are subjected to internal pressure of less than 150lbs. per square inch, and of riveted steel mains for higher pressures. the greatest pressure on any siphon is approximately 250 lbs. per square inch.

At the head of every siphon on each cast iron main is an automatic self-closing door, which is brought into operation by a mechanical device activated by the increased velocity of the water which results from a burst pipe. The automatic operation of these doors limits the amount of water which escapes from the fractured valves and are placed on each main at intervals of about one mile. Automatic air valves installed close to and on either side of each sluice valve and at every summit and marked change of gradient in the mains, provide for evacuation of air when the mains are being filled, and for ingress of air when they are emptied. Nine-inch washout valves and scour pipes are placed at every depression. Close to and on either side of each sluice valve is an inspection manhole.

The two mains were originally completely independent of each other, with the result that when a failure occured on one pipe on any of the eleven siphons, the capacity of the aqueduct was halved. In the early days of the life of the aqueduct this was not a particularly serious matter, as the combined capacity of the two mains was greater than the demand, so that the depletion of water in the storage reservoir at Frankley resulting from the failure could be made good after the repair had been completed. As the demand approached the combined capacity of the two mains, so the effect of a failure, with the consequent temporary loss of 50% of the capacity of the aqueduct, became more serious.

One obvious method of meeting the situation was to lay a third main but this could not be done because at the time when the position became critical the country was engaged in the Great War of 1914-18. Necessity gave birth to the hydraulic device of cross-connecting the two mains, on either side of the main valves. If, before the cross connections were installed, a burst occurred, the whole main was out of action. After cross-connecting, a burst at the same place put out of action only the length between two sets of valve and the loss of delivery capacity was thereby much reduced. To give an example in figures, the combined capacity of both mains was at one time 24 million gallons a day, so that, before cross-connecting, a burst main reduced the capacity to 12 million gallons daily. After cross-connecting, a similar burst reduced the combined capacity from 18 to 22 million gallons daily, dependent upon the position of the burst. This principle was applied when it became necessary to lay the third main on the aqueduct to meet the increasing consumption in the area of supply. Sections of the main were laid from time to time at different points along the aqueduct and the work was planned in such a way that by means of cross-connections to the existing mains the combined capacity was increased at a rate sufficient to provide a margin above the current requirements. This practice was obviously far more economical than the original intention to lay a third main as one continuous operation throughout the whole of the siphon sections of the aqueduct.

The first lengths of third main consisted of 60 inch internal diameter reinforced concrete pipes made on the site of the work, but only about two miles of these pipes were used, and the type of pipe subsequently adopted was a 15 foot long process with a one inch thickness of fine concrete and surrounded in the trench with four to six inches of concrete. As the frequent failures of the original 42 inch mains had occurred at certain points on the siphon sections where the pressure was high, they were strengthened by surrounding them with concrete at the same time as the third main was laid in their vicinity.

With the completion of the third main in 1939 the capacity of the aqueduct was increased to about $55^1/_2$ million gallons a day. In order to keep pace with the increasing demand, a fourth main, also of 60 inch diameter, was commenced in 1949 and completed in 1961. With its completion the capacity of the aqueduct was increased to 75 million gallons a day, equivalent to the approximate yield of the Elan Valley Works.

The siphon sections of the aqueduct cross 26 rivers and streams, special bridges being required at some places to carry the pipes.

The rainfall in 1895 was 53.47.

A report from the minutes on Wednesday May 20th 1903, told of 28" rain falling in three months and $5^1/_2$" of rain in one week. Only 15 days in the past year did the rain gauge show no rain. From February 1st to the beginning of May not one single day was fine.

CHAPTER NINE

The Walk

THE small town of Rhayader is the gateway to the Elan Valley dams. It has changed little over the past 50 years keeping many of its family-run businesses. Supermarkets are non-existent hence local businesses are able to give customers a personal service which is part and parcel of village life.

Rhayader has a history to offer those who are interested and much can be gained by speaking to local residents.

The Elan Valley dams brought visitors to the valley way back in the late 1800s and has continued to do so. Even in the middle of winter visitors can still enjoy the beauty and wonder of the valley.

Leaving the town of Rhayader via West Street on the right is Station Street (at the end you will find the remains of the old station). Just prior to the old road to Aberystwyth, which is again on the right, you can see the remains of the railway embankment which carried the railway built specially for carrying materials for the building of the dams.

The railway track is on the left when travelling along the road to the valley and also on the left-hand side you will see the old Baptist Chapel which replaced the one that was submerged.

The graves of the Lloyd family are in this graveyard. They farmed at Grove Farm. There was Richard Lloyd, who died in 1872, and he is buried with his two granddaughters, Ann aged two years and Enid aged three weeks. Then there is Richard Lloyd and his wife Ann who died in 1874; her husband died a year later. Elizabeth Lloyd died in 1894 aged 23 years. These were exhumed from the old chapel and re-interred at the present site. Altogether there were 60 bodies exhumed. The old chapel was used as a recreation centre by the engineers from Pen-y-Garreg house until the valley was flooded.

Approaching the Visitors' Centre is a Red Wind Sock which gives the wind direction should there be a chlorine escape. On this high ground above the road are the sand filters which clean the water etc. before it begins its journey to Birmingham 74 miles away.

Entrance to the village is by a bridge which crosses the River Elan. There is also the remains of the suspension bridge which cost £19.6s.7d. to build and was in use until the late 1960s.

The present village was built by Lovetts of Wolverhampton replacing the wooden huts etc that once housed the workmen and their families.

Most of the houses in the village today are occupied by employees of Welsh Water. One house in particular, Caban View, has changed little since it was first occupied by the Chitty family, Mr Chitty retiring in 1933 having worked in the valley from the commencement of the building of the dams in 1893. Part of the house was used as a village shop and accommodation was also offered to visitors holidaymaking and paying guests.

Little has changed since those days. Caban View is no longer a shop but accommodation is still available. There also remains the coach house, stable, hay loft and pig sty in the garden.

The Visitors' Centre was previously the workshops consisting of a fitters' shop, stores and blacksmith's shop which have been converted to an exhibition hall. There is also a small cinema and a café which serves tea, cakes and sandwiches, etc. The Ranger' office is also housed in this building. Walks of various distances are available which incorporate bird watching, etc.

Of the four dams, Caban Coch is the lowest in the chain being 122 feet high and 610 feet long, holding 7,815 million gallons of water. The water stored in this dam does not supply Birmingham. The Birmingham Corporation Water Act of 1892 made provision to keep the balance of the Wye correct (the Elan River being a tributary). Before the water from this dam joins the River Wye it is passed through two power houses, one on each bank of the river. Power is generated by turbines driven by the compensation water, the electricity being used on the Elan works.

Opposite Caban Dam there is a car park where you can see the remains of where the dynamite was placed for blasting the rock away which was then given to the stonemason to cut and shape before it was placed on the face of the dam. About 200 yards further along this reservoir on the far side can be seen a small inlet in which was built a small dam for storing water for villagers' use while the larger dams were being constructed. When the work was completed it was no longer required but during the 1939-45 war it served a very useful purpose. It was here that Barnes Wallis spent many hours carrying out experiments with his bouncing bomb invention. When he was satisfied with the results he then went to Derwentwater to complete his tests with aircraft until he was ready for the final bombing raid on the Mohne Dams in the Ruhr.

The confluence of the rivers Claerwen and Elan takes place in the Caban Reservoir. The site of Nantgwilt House is under this water. It was here that the poet Shelley lived for a short while with his wife Harriet. Nothing now remains except the garden walls which can be seen when the water level is low. The house is believed to have inspired the novelist Francis Brett Young to write 'The House Under the Water'.

Mr Yourdi also lived at Nantgwilt while he was resident engineer-in-charge of the building of the dams.

One-and-a-half miles further upstream from Caban is the Garreg Ddu viaduct. This is a road carried on arches which are supported by piers on a dam below. This is sometimes affectionately known as the Submerged Dam.

It was in this area that the Elan village was situated before being submerged. Most of the buildings were rebuilt except for two large houses, Nantgwilt and Cwm Elan.

It is worthwhile making time to visit Nantgwilt Church which is lit by oil lamps and candles there being no electricity in the valley.

The Foel Tower, which is on the opposite side from the church, controls the water which goes to Birmingham. It houses the gear controlling the flow through the Foel tunnel to the Treatment Works. Water from the Claerwen Dam also travels by tunnel discharging her water into this reservoir. This happens during the summer months when the water levels fall low.

Mr Mansergh, when designing the dams, realised that there would be problems during summer months hence the reason for the submerged dam.

As previously mentioned the balance of the Wye has to be maintained to a certain level. When the water level falls to 40 feet the submerged dam comes into view. It is at this point that the Foel Tower controls sufficient water into Caban Dam to maintain the level of the River Wye and at the same time sending water to Birmingham.

This wonderful design ensures that Birmingham receives its supply of water during dry weather when many other towns and cities have hosepipe bans, etc.

Still following the bed of the old railway towards Pen-y-Garreg on the far side of the reservoir there is a small valley in which can be seen the remains of the old lead mines. Soon after this the road approaches an arched masonry bridge. The Birmingham Water Department won an award for the rebuilding of this bridge in 1965. Coaches bringing visitors to the valley had to turn around here the bridge being too narrow for them to cross over. It was decided to remove half of the bridge, widening it and replacing the old arch. Each stone had a number on it so that they could be replaced in their original position. These numbers can still be seen. Crossing the bridge you get a wonderful view of the Pen-y-Garreg dam which is the smallest but the one that attracts most attention. This dam is 128 feet high and 500 feet long and holds enough water to supply Birmingham for 15 days. The tower in the middle is reached by a narrow gallery which runs through the centre. Light comes in through openings in the face of the dam. When water is pouring over the crest it is like looking through a lace curtain.

It is possible to walk along the far side following the bed of the old railway. All the bridges on this railway were of a wooden construction except one which was built of stone. It can be clearly seen from the road. Just beyond the stone bridge is where Mr Hughes re-erected his shop which was used as a chapel until 1965. The pews from this chapel were removed to Rhayader Church. Further on still is Devils Gulch. It was at this point builders of the railway encountered one or two snags: rick had to be blasted away and also there was a curve in the railway line. This can also be viewed from the road.

After passing the reservoir look for a white marker board high up on the hillside to your right. In the 1960s there was a proposal to build a larger dam which would have meant much of the area going under water and the dam itself would have been larger than the Careba Dam. The project would have been very expensive and was abandoned. The top dam, Craig Goch, was the last in the chain, this being built on a curve.

Continuing higher up onto the plateau you will find very few trees. Vegetation provides enough food for the sheep who graze there. Altogether there are about 40,000 sheep on these hills which are rounded up sometimes on horseback or buggy. Every sheep farmer must keep back at least 200 of his sheep when sending the rest to market so that when new arrivals join the flock they can be taught the sheep walks etc. You also pass a farm on your right by the name of Hirnant's Farm. This has the most expensive telephone installation in Wales if not the whole of the British Isles. At the finger post the left fork takes you to Aberystwyth and the right to Rhayader this is the one you take. Quite often the red kite can be seen which is a most beautiful bird. In recent years two wind farms have been erected in the area whether they are worthwhile or not is open to debate.

Carrying on along this road you get beautiful views before coming to a T-junction, by taking the right-hand direction you return to the valley, look for a red brick building on your right-hand side situated in the trees. It has what seems to be three windows in one wall this is part of the Birmingham water supply. Approaching the Visitors' Centre you will notice a small brick building opposite the old suspension bridge. It was here every Saturday at midday that navvies would collect their wages from Mr Chitty. This building holds many memories and would have a fund of stories to tell of people engaged in the building of the dams.

I have come to the end of my story which I hope you have enjoyed and should you visit the valley on some future occasion I hope that what I have written will have convinced you what a wonderful place it is and also given you an insight about the men who had the vision and foresight, and determination, to build the dams. Last, but not least, to those wonderful navvies.

An excerpt from an old local guide book:
As you pass over the lower flank of Clee Hill going towards Hopton Wafers you suddenly come upon an astonishing building standing in a field beside the road. The bizarre, brick-Gothic, castellated shape is so hideous that it exercises a kind of mesmerizing fascination to which you reluctantly succumb, especially when you discover that it has a touch of

poetry about it after all; for this is a water station, one of the vital links ensuring that water from the romantic Elan valley lakes, gleaming between their wild and craggy Welsh mountains, is safely piped to unlovely and unromantic baths and basins in Birmingham.

Travelling from Birmingham to the Valley by road.

The journey to the Valley from Birmingham is most picturesque and at several points along the route the Aqueduct can be seen.

One travels through the Village of Brampton Bryan. This village has no Public House, the reason being during the construction of the Aqueduct two Irish navvies had a fight to the death in the Village Pub. Mr. Hartley the local squire and land owner closed the pub and then vowed that there would never be another in the Village.

A story is told that when the third pipe was being put in at Leintwardine trenches were dug by hand to a depth of 12-18 feet.

At about 4 a.m. each morning the local game-keeper would walk the length of the pipe-line and kill any deer that had fallen into the trench during the night. The man telling the story was asked if he worked as one of the navvies as he knew such a lot about the happening there. "No" was his reply "I was the game-keeper". This gentleman now lives in Alcester.

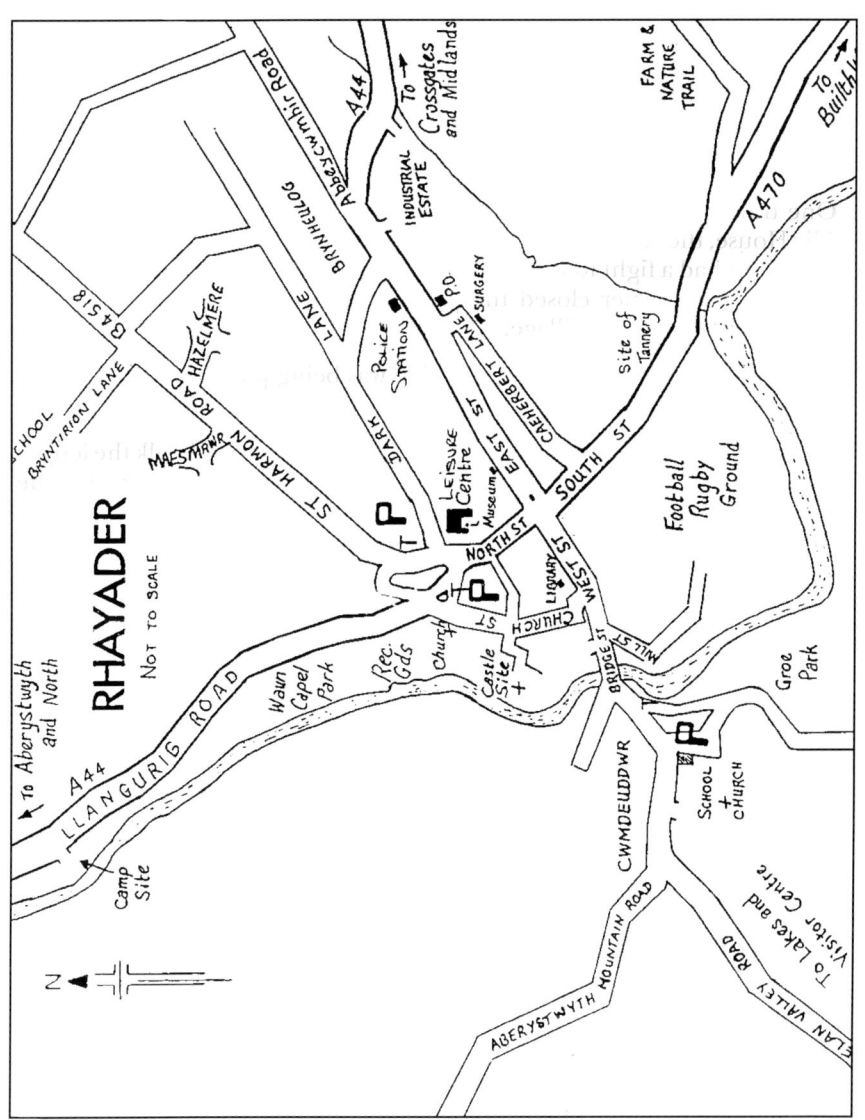

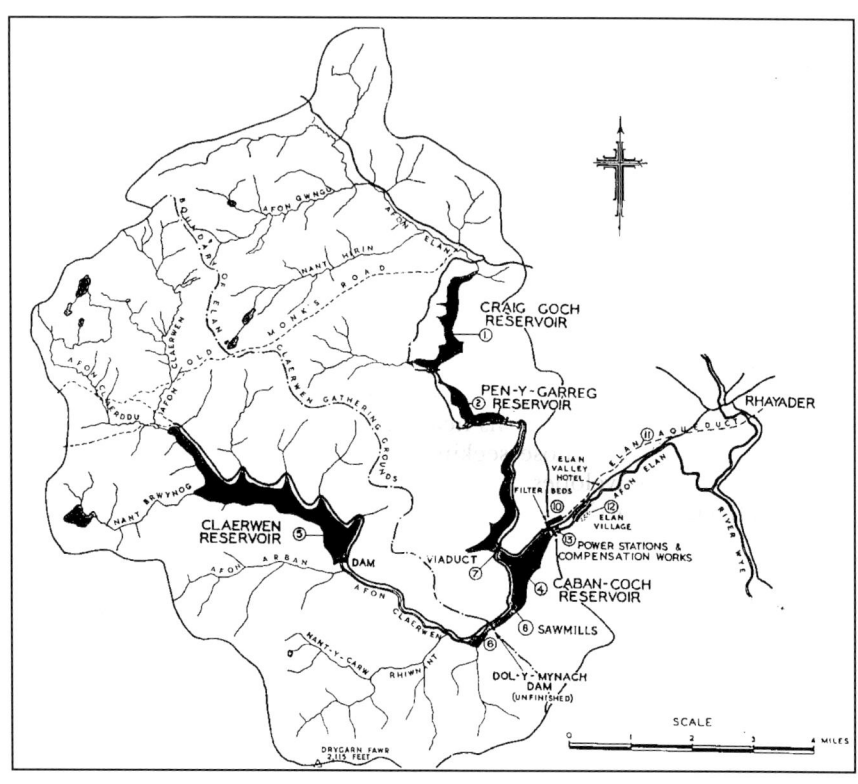

Translation of Welsh Names
Caban Goch   Red Cabin
Garreg-Ddu   White Rock
Pen-y-Garreg Above the Stone
Craig Goch   Red Rock

Contents	Height	Length	Top Water Level OD	Top Water Area	Contents
1. Caban Coch	122ft.	600ft.	822ft.	500 acres	8,000 m.gals.
2. Pen-y-Garreg	128ft.	500ft,	945ft,	124 acres	1,300 m. gals.
3. Craig-Goch	120ft.	120ft.	1,040ft.	217 acres	2,000 m. gals.

About the Author

Rita Morton was employed by Severn Trent Water Authority for nearly 25 years in a position she enjoyed immensely. It was her job to work with schools and other groups explaining the treatment works. This knowledge was gained from Works Superintendents who were experts in their field.

Birmingham Water Department recorded everything connected with their water supply. These records, from a young boy's application for a position with references from his headmaster, minister of his local church and a family friend, to those seeking positions of a higher grade, were kept in individual envelopes, some of them going back to the mid-1880s. This, in effect, was a short history of Birmingham and made interesting reading.

As plans were being made for privatisation it was decided to scrap most of this documentation and it all found its way into a skip for burning.

Mrs Morton was given permission to take whatever she wished. Time was short so she recovered what she could taking it away in sacks.

This book is the result of her searching through this material and logging all the relevant information she needed. She is now an authority on the Elan Valley and has lectured far and wide and at the same time met many interesting people from whom she has gained further information.

This she acknowledges in the book, the rest being her own research. Posters that were found in the skip were ironed out and reduced and copied at the Edgbaston Depot. These were left at Tame House and trust that they are being looked after.

There are several people she wishes to thank for help in enabling her to write this book.

A very special thanks to Anthony Biddle for his help in making this book possible.

To Richard and Gill Hughes of Rhayader for their dear friendship and hospitality and loan of photographs, etc.

To Susan Smith who right at the beginning offered her help and services in spite of her own busy life.

to Mr. A. Chitty and his daughter Rita who have given me help with regard to the family life in Elan Valley.

My thanks also to Mrs. J. Pite of Worcester who loaned to me family photographs and also gave me useful information about her grandfather Mr. Atkinson.

John Irish for the loan of valuable material.

To Esther Fernando for the loan of equipment in the early days of the book.

To Michael Bennett at the Elan Valley Works who was always willing to give his time and knowledge of the works. He leads a very busy life but always found time to help me.

Liza of Welsh Water for her help in many different ways.

To Tom Hemming and his daughter Eileen who have been wonderful in helping to get everything in order and sending the finished article to the printers.

And to all the groups I have spoken to in different parts of the country as far away as Jersey who have shown great interest and given me encouragement to complete the book.

To everyone, a very sincere thank you.